THE FALL OF YARMOUTH SUSPENSION BRIDGE:
A NORFOLK DISASTER

Further details of Poppyland Publishing titles can be found at
www.poppyland.co.uk
*where clicking on the 'Support and Resources' button
will lead to pages specially compiled to support this book*

Join us for more Norfolk and Suffolk stories and background at
www.facebook.com/poppylandpublishing

Dedicated to the memory of the 78 people who lost their lives on
2 May 1845. They are not forgotten.

The Fall of Yarmouth Suspension Bridge:
A Norfolk Disaster

by

Gareth H. H. Davies

Copyright © 2015 Gareth H. H. Davies

First published 2015 by Poppyland Publishing, Cromer, NR27 9AN
www.poppyland.co.uk

ISBN 978 1 909796 21 8

All rights reserved. No part of this publication may be reproduced, stored in a retrieval system or transmitted by any means, mechanical, photocopying, recording or otherwise, without the written permission of the publishers.

Designed and typeset in 10.5 on 13.5 pt Gilgamesh
Printed by Lightning Source

Picture credits

Author 10 top, 22, 24, 27, 29, 95, 97, 102
Author's collection 10 bottom, 11, 13, 16, 18, 19, 25, 28, 38, 39, 40-41, 50, 53, 54, 55, 57, 58, 59, 60, 62, 63, 66, 69, 71, 73, 81, 84, 93, 100
British Museum, Public Domain 12 (C.194.b.305-307), 78 (B20073 84)
Archant, Norfolk Museums Service 96
Stephen Craven, Creative Commons 47
© Victoria and Albert Museum, London 88

Contents

	Preface	6
	Acknowledgements and Glossary	7
	Introduction	9
1.	The Disaster	15
2.	The Victims	21
3.	The Cause of the Collapse	37
4.	Yarmouth Society	49
5.	The Building of the Bridge	57
6.	Road, Rail and Bridge	65
7.	Circus and Clown	77
8.	Legacy and Memory	93
	Appendix – List of Victims	107
	Appendix – Narrative, R. Cory, 1832	113
	Appendix — Miscellanies Relating to Yarmouth	117
	Index	121

Preface

Beside the waters of the River Bure in Great Yarmouth, Norfolk stands a shiny black memorial erected by the townspeople on 28th September 2013.

This memorial commemorates those who lost their lives in May 1845 in one of the most bizarre and tragic disasters this country has ever known. The memorial itself is the culmination of a personal mission to both honour those that died, and provide a permanent reminder of this incident in the history of the town. Julie Staff, who runs a deck-chair business on Great Yarmouth beach, became aware of the story in May 2011 and decided to raise money for the memorial.

> I didn't want to raise money from grants and funds like the Lottery, I wanted it to be raised by the people, so I took my campaign to the beach. Locals and holidaymakers alike got a deck chair and a history lesson! ... The erection of the memorial has now given people a place to pay their respects and remember the victims of the disaster, many of the descendants would never have even learned of the disaster and the fate that befell their relatives had I not started my journey. I am pleased that my project has ensured that the memory of those that died in the bridge tragedy has not been forgotten and the town now has a lasting memorial to the events of that day[1].

My own "journey" with regard to this story began in 1983 when looking for a local history topic that would engage my history class. Working with a colleague from another school I produced a document pack with teaching material for use in local schools. Mrs Staff's laudable effort spurred me to revisit the work done then.

This book is an attempt, in small part, to provide the background story to those who might come across this memorial in the future and want to know more. Since 1845, the story has ebbed in and flowed out of the collective memory of the town being re-told in newspapers and magazines, and more recently on television and the internet. These publications have done little but use the story as an anecdote to interest their readers or audience. Where the event is mentioned in histories of the town it is as a tragic footnote. The memorial deserves more than that if it is not to become just another curiosity for future generations.

Gareth H. H. Davies

1 Door to Door – Broadland Housing Association Tenants Magazine, Spring 2014 p. 13. Available at https://www.facebook.com/groups/350701568294397/permalink/758528467511703/ (last accessed 26 January 2015).

Acknowledgements

I have not written this volume without the help and encouragement of others. My first encounter with the story was when I was working with a teacher colleague, Chris Tooth. We met on a course for teachers entitled "Teaching Historical Skills" in 1981/2 and as part of that course produced a document pack on Sarah Martin and Yarmouth Prison. It began a fruitful working relationship and our second local history pack was on the story of the Suspension Bridge, which was distributed to local schools in 1983. Much of the initial research was therefore done in 1982 and 1983 and our intention then was to ask our students questions rather than have our own opinions on the event. As a result many questions remained unanswered in my head and I always felt that there was still more research to do. In 1995, on the 150th anniversary of the disaster, I was encouraged to give a talk on the disaster and naively believed that this would provide the impetus to put pen to paper. This was not so, and only when I heard of Julie Staff's efforts to raise money for a memorial to the victims and the fact that her dedication proved successful did I gain enough courage to turn asking questions into an attempt to answer them. In the intervening thirty years access to source material has become significantly easier particularly with the digitisation of newspapers and their placement online.

A special thanks must be given firstly to the staff at Great Yarmouth Library of thirty years ago, which I apologise I can no longer name, and secondly, my wife, Janet who has acted as a careful and insightful reader for this book. Finally, I must thank Peter Stibbons at Poppyland Publishing for taking on the work, your friendship and encouragement has been invaluable.

Gareth H. H. Davies, 2015

Glossary

Havel and sleas types of banners hung on poles and carried at parades and marches.

Last (scheepslast) a dutch measurement of fish literally meaning a "load". A last was the equivalent of 120 cubic feet (3.398m^3) of shipping space.

Pantaloon The Harlequinade evolved in English pantomime from the traditional Venetian entertainment and consisted of four main characters — Harlequin, Columbine, Clown and Pantaloon. Pantaloon was the father of the Columbine, an old man that the other characters made a fool of, the Clown most of all.

Introduction

1845 saw the publication of Benjamin Disraeli's novel *Sybil* or *The Two Nations* in which this ambitious author and politician was beginning to develop his political philosophy of Social Toryism. In the novel, Disraeli warned that Britain would become divided into "two nations", of the rich and poor, as a result of increased industrialisation and inequality in society. Disraeli's "one-nation" conservatism proposed a paternalistic society, which emphasised the importance of social obligation rather than individualism. While Disraeli's novel was the rhetoric, the story of the collapse of Great Yarmouth's suspension bridge, in that same year, provides us with the reality — a community coping with a traumatic event, while struggling at a time of increasing economic, social and political change.

History has been described as a continuous process in which events either exemplify the past or provide dramatic interventions, or turning points, in its course. However, continuity and change are rarely as clear-cut as this, whether describing the past or the present. One thing is certain, in order to consider the past, we must have some understanding of the time. People's actions are governed by their milieu. The events and actions of those found in this book therefore need to be put in the context of their time. Here are some pointers:

The growth of population. The population of Great Yarmouth, like that generally of Britain, showed an inexorable increase throughout the 19th century. It was in the decade of the Napoleonic Wars (1801-1811) and in the period this book is concerned with (1841-1851), that Yarmouth saw its greatest increases in population, 21% and 28% respectively.

These dramatic spurts in population growth tell us something about the economic development of the town and both the need for, and the result of, infrastructure to support a growing economy.

The consequences of the French Wars. The end of the wars brought unemployment, trade depression due to increased foreign competition, and continued poverty to many parts of the country and Yarmouth was no more immune than the rest of Britain. The town had been an important military base during the wars with a naval hospital and armoury being built in the town. Although, the fisheries were in a parlous state generally, Yarmouth's economy was stagnant rather than depressed[1] and often supported agricultural labourers during the herring season as they were taken on to assist with pulling in and handling the nets.

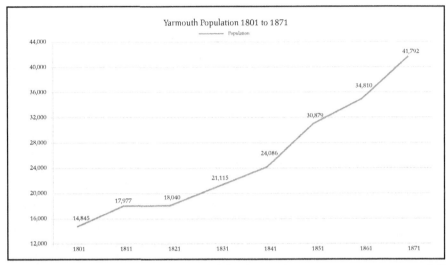

Yarmouth Population 1801 to 1871

The Transport Revolution. The need for better transport systems was vital if Yarmouth was to take the economic opportunities presented to the town during this period. In the late 1820s, the suspension bridge itself was a reaction to this need, as was the subsequent construction of a turnpike road across the marshes to Acle. In 1844, Norfolk's first railway linking Yarmouth and Norwich was built, and this was to change everything.

Social and Political Theory. The two dominant social and political theories of the period were Utilitarianism and Political Economy. Utilitarianism, the philosophy of Jeremy Bentham, argued that human nature was motivated by self-interest, or more exactly, the balance between pleasure and pain. The principle of utility was that every action was governed by the "pursuit of happiness". Bentham developed a method of calculating the value of pleasures and pains, and one of these measures was the extent or the number of people affected by an action. Hence, "The business of government is to promote the happiness of the society, by punishing and rewarding ... In proportion as an act tends to disturb that

Brandon Station, 1845

happiness, in proportion as the tendency of it is pernicious, will be the demand it creates for punishment."

Political Economy was categorised by the works of Adam Smith. In his book, *The Wealth of Nations*, he argued that society has moved through a number of phases — hunting, nomadic agriculture, then settled farming, and finally a system of commercial interdependence. Smith suggested that in each of these phases society had developed institutions appropriate to that phase. Each phase was governed by the interplay of self-interest and morality. According to Smith, commercial interdependence required the government to pursue a policy of laissez faire (let alone) with the economy. Under *laissez faire* systems, individuals, acting in their own self-interest, would tend to dedicate themselves to those economic activities that brought them the greatest reward in terms of income, be it in the form of wages, rent, or profit. Smith showed that by giving themselves to such highly rewarding economic activities, in their own self-interest, people would also be maximising the economic well-being of society.

The Evangelical Movement. By this period the religious movement called the "Evangelical Revival" was firmly established among the upper and middle classes. It demanded stricter conduct by all and a belief that Christian ideals could save all but the most depraved in society. Redemption through repentance and education in Christian teaching was seen as the main focus of this revival.

The Condition of the Poor. The condition of the working poor became an increasing issue for 19th century society as the population grew and industrialisation took hold.

Society became more complex and there was a growing realisation that previous ways of managing society did not necessarily meet the needs of the time. This manifested itself in opposing social theories, political groupings and protest movements, as well as practical legislation and charity. Yarmouth society reflected these national tensions as keenly as other towns and cities.

A contemporary image of the dwellings of Manchester operatives during the cotton famine of 1862

Forms of entertainment. The first half of the 19th century saw the rapid development of popular entertainment to a wider audience. Indeed, Great Yarmouth may have had the earliest documented purpose-built theatre in England. The Game-Place House is first mentioned in 1539 when the bailiffs and chamberlain leased it to Robert Coppyng, almost 40 years before the first London playhouse[2]. In 1778 a purpose-built theatre was erected on what is now Theatre Plain.

Astley's amphitheatre, 1808, as shown in A Microcosm of London.

Circuses began in 1768 with Philip Astley in London and by the 1820s there were many touring troupes. The first recorded instance of a circus in Yarmouth was in April 1820, when the New Olympic Circus set up on Theatre Plain[3].

One of the most important public entertainment events was the Yarmouth Annual Water Frolic at the end of July in which river yachts competed against each other and rowing matches were conducted in the evening, both for a silver cup[4]. This, together with the Yarmouth Races, held on the South Denes over two days in late summer, were the highlights of the town's year and attracted many stalls, hawkers and fairground entertainers. As John Preston, in his Picture of Yarmouth, published in 1819, put it:

> The polite amusements of the Assembly Room, Theatre, Bath Room, Concerts, and annual Races, render the residence of visitors whether for health or pleasure extremely agreeable: those who are fond of fishing or sailing, may indulge themselves satisfactorily in the vicinity of the town. Besides the above first class of amusements, there are others of no ordinary attraction during the summer, and much praise is due to the proprietors of Vauxhall and Apollo Gardens, for the expence they have recently been at, in order to render these much improved places worthy of patronage of the public, and where the Bowling Greens are kept in the highest order.[5]

By the beginning of 1845, Yarmouth was looking to the future. The 1844 herring season had been good[6] and the railway had made a significant difference to trade in the town. In October 1844, this optimism was reflected in the call for a public meeting to discuss how even better communications with London and improvements made to the harbour could be achieved, so that Yarmouth could

The Yarmouth Theatre, 1819, from Preston's Picture of Yarmouth.

maintain its pre-eminence and extend its trade[7]. Yarmouth was keen to ensure that it did not lose out to others and was ambitious to grasp the opportunities that were presenting themselves. The key to this golden future was transport and infrastructure.

[1] *Norfolk Chronicle* — 9 November 1833, p.4. —
"The Report of the Committee to inquire into the British Channel Fisheries has been published. The Committee regret they have to report that the fisheries and various interests connected with them are in a deeply depressed state, that they appear to have been gradually sinking since the peace of 1815, and more rapidly, during the last nine or ten years; and that the capital employed does not yield a profitable return, while the number of vessels and boats, as well as of men and boys employed, is much diminished, and the fishermen and their families, who formerly were maintained by their industry, and enabled to pay rates and taxes, are now in a greater or less degree dependent upon the poor rates for support. The observations of the Committee are continued to such places on the coast as lie between Yarmouth and Cornwall. The Committee ascribe this falling off to the following causes, which they consider immediately susceptible of remedy, viz. a large quantity of foreign caught fish, illegally imported and sold in the London market; and a great decrease and comparative scarcity of fish in the Channel. ... The Committee recommend, that fish carts should be exempt from the payment of tolls. ...

THE YARMOUTH FISHERY. — The fisheries of Yarmouth appear to your Committee to be in a comparative prosperous state, but suffering much in the herring fishery from the heavy duty levied at Naples upon herrings exported thither, being 15s per barrel, which for the last three or four years had amounted to a prohibition; ...

The fishermen of Yarmouth also complain of the injury they suffer, particularly in the herring and mackerel seasons, from competition with the French fishermen, and from the illegal importation of those fish, especially mackerel, when caught by foreigners, not being prevented; this they earnestly request may be done ...

The fishing vessels belonging to Yarmouth are stated to be about 100 sail, averaging from 40 to 50 tons each, besides 50 or 60 vessels annually hired from Yorkshire during the herring season; and the number of men employed at sea are computed at between 4000 and 5000, besides a great number to whom employment is also given on shore. The capital embarked is estimated at about 250,000l."

[2] *Norfolk Record Office Information Leaflet 27* available at http://www.archives.norfolk.gov.uk/view/NCC098513 (last accessed 21 February 2015).

[3] *Norfolk Chronicle* — 1 Apr 1820, p.3. — "Ladies and Gentlemen of YARMOUTH and vicinity are most respectfully informed, that Mr. COOKE has fitted up, on the THEATRE PLAIN, an elegant and commodious CIRCUS, which he intends Opening on Friday, April 7th, 1820.".

[4] The water frolic was held in late July and was traditionally concerned with the meeting of the mayors of Norwich and Yarmouth in their state barges. Accompanied by their Water Bailiffs, they met on the River Yare at Hadley Cross near their civic and parish boundaries. They then "beat the bounds" to Breydon Water ensuring that all boundary posts were in place. According to James Stark's "Rivers of Norfolk " (1834) they were accompanied by a regatta of boats laden with revellers and conduct "a fair afloat". The frolic was renowned for unruly behaviour, and, for this reason, the tradition of making regatta day a civic holiday in Yarmouth was abandoned in 1834.

[5] Preston, J (1819), *The Picture of Yarmouth: Being A Compendious History and Description of all the Public Establishments within that Borough* available at https://books.google.co.uk/books/download/The_Picture_of_Yarmouth.pdf (last accessed 25 March 2015).

[6] *Bury and Norwich Post* — 23 October 1844, p.3. — "The boats employed in the Herring Fishery are, we are happy to say doing much better than they were last year. The quantity of herrings shipped to the London market on one day last week from Yarmouth and Lowestoft amounted to the enormous quantity of 91 lasts."

[7] *Norfolk Chronicle* — 26 October 1844, p.3. — "A requisition to the Mayor is now in the course of signature, requesting him to convene a public meeting for the purpose of taking into its consideration the best method of obtaining a communication from Yarmouth, through Diss and Colchester, to London — of promoting increased facilities for the transmission of goods — for improving our harbour, and for extending the trade of the town; with a view to prevent the diversion of the trade now enjoyed by it, to other ports; and with the hope to extend it. As this subject is of vital interest to every merchant and tradesman in Yarmouth, we hope the above meeting will be attended consistently with its importance."

1

The Disaster

"Oh Mercy what a scene was there,
The waters leapt up into air,
And mangled, in their edding strife
Forms midway between death and life!"
The Parting and the Meeting
Or The Burial of Yarmouth Bridge
James Stuart Vaughan, 1847

"Never has it been our duty to relate so woeful a calamity in our district as that which we have this week to lay before our readers", so the *Bury and Norwich Post* reported on 6 May 1845.[1] On Friday, 2 May, entertainment turned to disaster when crowds gathered on the suspension bridge spanning the Bure had been plunged into the cold waters of the river. In all, 78 lost their lives, many women and children. Among the shocking disasters of the Victorian age, this became marked as unique in circumstance and horror and had a profound effect, not only on the town of Yarmouth but throughout the country.

It was in May 1845 that Cooke's Royal Circus was to entertain the crowds at Yarmouth. A family business, Cooke's Circus was well known throughout the country and abroad specialising in equestrian extravaganzas. The proprietor, Mr William Cooke, was renowned as one of the best horse trainers in the business and acts included his daughter, Kate Cooke, billed as "The Infant Wonder at three and a half years old".[2] Like all circuses there were other entertainers – tumblers, slack and tight rope walkers and clowns. The resident clowns that season were Mr. Nelson and Mr. Swann. It was Nelson, described as "principal low comedian" and "modern day Yorick", who was to draw the crowds on that day and sail to the suspension bridge in a wash tub pulled by four geese to achieve publicity for his benefit night at the circus.

The wash tub stunt had been performed many times before, both by Nelson and other circus clowns. The plan was that Nelson would set off from Yarmouth Bridge at five thirty in the afternoon and, helped by the tide, "sail" up the

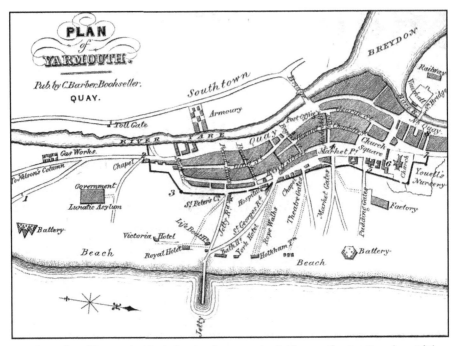

Barber's plan of Yarmouth, 1846, showing the suspension bridge over the river Bure (top right).

River Yare and into the Bure, finally alighting at the Vauxhall Gardens next to the suspension bridge. Yet everything did not proceed as planned, for strong currents drew the tub, and the boat pulling it by a weighted rope, further than expected up the Yare and into the mouth of Breydon Water. The delay heightened the expectation and excitement of the crowd in their hundreds lining both sides of the river. In excess of 400 had gathered on the suspension bridge jostling for the best view. On hearing the shout, "Here come the geese", they moved to the south side of the bridge en masse[3]. One observer later commented, "the bridge once convex roadway had flattened". With hardly a warning, the excited crowd was projected into the river as one side of the south suspending chain gave way.

Many horrific and heroic tales were told of the next few seconds and minutes as men, women and children became an entangled mass with wood and chain. "One man fell across a piece of iron with his head just above the water, he reached to two girls who were struggling for life, and a third caught hold of his collar. All four were saved". "Eliza, the daughter of Mr James Borking, a dyer, aged 12 whose sister was drowned survived when she got hold of a man's leg and pulled her out." While the son of Mr Jay, the baker in White Lion Gates, said that when he was under the water, the people looked as if they were hugging each other. "He could see them quite perfectly."

The *Illustrated London News* reported that:

Those who witnessed it asserted that not a scream was heard, nor a sound emitted from the unfortunate victims. A fearful splash and a few gurgling struggles, only recognised the spot which had swallowed such a mass of human life. Some few men hanging by the broken chains were earnestly entreated to maintain their hold, but it was soon observed that, in consequence of the obstruction of the stream by the fallen bridge, and the human bodies below, that the advancing tide would soon bury them even from sight. Every boat was immediately in requisition, and as many as twenty-five were soon on the spot, and rendered active and gallant service. The scene at this moment beggars description – husbands and wives, parents and children, were excited with the deepest anxiety. The efforts to save the victims were noble and praiseworthy. …

The various incidents of the tragic scene are some of them very touching. The children, many of them found with their heads fast in the railings (which on bursting the chains, lapped over into the water), doubtless fixed in that position in their anxiety to feast their eyes on the expectant sight. It was with difficulty that they were extricated, and there were some discovered with their heads smashed to pieces by the falling iron-work.

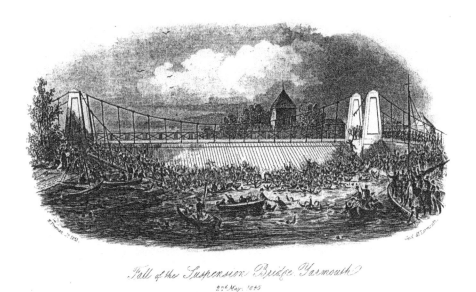

The fall of the bridge as shown in Freeman's print, published in Yarmouth.

Sarah Bammant's eye-witness account some eighty years later gives us a flavour of that afternoon:

> I was then about ten years old, and my aunt had asked my mother to let my two sisters and me go and have tea with her, and she would take us around to the Suspension Bridge to see the theatre-man come up river in his tub. We found many people there, and many children, and we crossed through the crowd standing on the bridge to the Runham Vauxhall side, where there were not so many people, and we could see better. My sister Harriet was only a little thing of six and could not see at all, so a lady said she would look after her and took her just on to the bridge where she could see better. My aunt, my sister, and I stood just on the Vauxhall side of the bridge, leaving Harriet with the lady a step or two away.

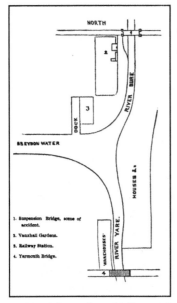

A plan of the site from the Illustrated London News May 10th 1845.

Someone shouted, "Here he comes!" and the geese could just be seen coming around the bend in the river. People rushed to the side of the bridge, and all at once there was a loud alarming crack. I turned around with a cry, "Where's Harriet?" I just clutched hold of her and pulled her from the bridge to me standing on the edge of the riverside. This quick almost involuntary act, which doubtless saved her life, swung me round so my back was to the bridge and I did not see the actual fall of the structure. But when I looked again it was dreadful. One side of the bridge had given way, and was hanging down perpendicularly, and all the people on it were shot by what was like a big chute into the water. The lady who had taken my sister had disappeared. The screaming was terrible, crowds struggling in the water pulling each other down, and I saw hands pushing up above the surface. Then boats began rushing out rescuing all they could. We stood there crying and frightened, having to wait a long time before we could be taken back to the Yarmouth side as more boats were coming and getting people out of the water, and vehicles were arriving with blankets.

When we had been taken over my aunt said she must take us home to allay mother's anxiety, and on the way we met such lots of people madly running and asking, "Is the bridge down?". We lived in Adam and Eve Garden then so we had a long way to go and it was very terrible. The town seemed stunned I think over 70 drowned, mostly the children of tradespeople who had been taken down to see the sight. Some parents lost all their children. Ah! It was a bad day for Yarmouth.[4]

As the evening drew on bodies were picked out of the water and the half-dead and dead were taken to the Norwich Arms and other public houses in the vicinity. Barrels of hot water were brought from Lacon's brewery and the victims immersed and rubbed to try and revive them. The tap-room, kitchen and other rooms became a field hospital and the stables the temporary morgue. By 11 o'clock at night the body count had supposedly reached 73, with only ten to a dozen revived. The bodies were removed as relatives identified them and at twelve o'clock only one was unclaimed. In the meantime nets had been laid across the river with the expectation that, with the turn of the tide at 2am, other bodies would be washed back in the tow. On Saturday seven bodies were supposedly taken to the Norwich Arms including Ann Becket and Reeder Thurston Balls. As the weekend commenced others were reported missing. As a result, on Saturday, the Coroner issued a notice that anyone that had lost relatives register their names at the Police Office to ascertain whether they were found or still missing. By Monday evening only Louisa Utting and James Seaman Buck remained on this list as feared drowned. By this time the bridge had been removed to reveal that no other bodies lay under the iron-work.

On Saturday morning at 12 o'clock W.S. Ferrier, the borough coroner convened an inquest at the Church Hall. He swore in 19 men[5] of the borough saying, "There has been vast excitement produced, and I consider it very important that a very intelligent Jury should be summoned. ... you will have to inquire not merely *how* these individuals came to their deaths, you will have to inquire into *the cause* of this vast sacrifice of life."[6] Dr. Ferrier had suggested they take two to three cases that Saturday and then adjourn until Monday. However, many of the jury felt that they needed to move as quickly as possible given that the inquest could take some time. Since the grieving families had taken the deceased to their homes, it became necessary that the jury visit each house in turn. The jury began

A contemporary picture showing the removal of the bridge on Monday 5th May 1845.

their task by going to the Livingstone home which was above their shop at the corner of the market place and King Street. William Livingstone was a draper and his wife Martha had taken their three children Matilda, aged 7, Joseph, aged 6 and William aged 4 to see the spectacle the previous evening. Martha and William were the only two to survive. On their return, Samuel Palmer, the foreman of the jury also brought up the case of the body still lying at the Norwich Arms, which had not been identified the night before. Now it was shown, by relatives who had come to collect the body, to be Isaac Bradberry, a Norwich shoemaker. Isaac was one of two victims from Norwich and lived in King Street with his parents, John and Anne and their large family[7]. While the jury were viewing this body, Mr Bale, the officer of the court, was instructed to draw up a list of those dead and provide a route that would take the jurors around the town. The jury took seven hours to visit the homes of the victims and it was not until 9 o'clock that evening did they finish their journey[8]. The task, given the grief of the victims' loved ones, was extremely distressing, but these ratepayers[9] were little prepared to meet head-on the poverty and destitution among the poor of their town. As one juror put it, "To attempt to describe the state of many of the families is together beyond our power. We have beheld scenes of misery which would soften the heart of the most insensitive, and which caused intense pain to those whose mournful duty it was to enter their abodes".[10]

[1] *Bury and Norwich Post* — 7 May 1845, p.3.

[2] *Norfolk Mercury* — 11 January 1845 — Advertisement for Cooke's Royal Circus at the Amphitheatre, Victoria Gardens, Norwich.

[3] *Illustrated London News* — May 10 1845, p.297. — This text is probably from the Norfolk News and re-used. It appears in a fuller account in the 19th issue on May 10 1845.

[4] *Yarmouth Mercury* — 1929 available at: https://www.facebook.com/media/set/?set=a.377978655651256.1073741834.241973662585090 (last accessed 8 February 2015).

[5] *Norfolk News* — 10 May 1845, issue 19, p.3. — Samuel T. Palmer, Esq., foreman, John Norman, John Orfeur, James Pratt, John Fenn, Mr. Gurwood, B. Palmer, William Smith, William Spillings, Charles Barber, Charles Wolverton, Joseph Davy, James Emms, William Hallett, Mark Blowers, J. Eastmore Lawes, John Stagg, Edward Garrod, Thomas Davy.

[6] Ibid.

[7] 1841 England Census for Norwich, HO107/769/6 30.

[8] *Illustrated London News* — May 10 1845, p.297.

[9] Only men, who were ratepayers, aged between 21 and 60 were eligible to make up a coroner's, or indeed trial, jury. The foreman was usually a gentleman of the rank of esquire. In the case of this jury, the foreman, Samuel Thurell Palmer Esq was a solicitor in the town, while the rest of the jury was made up of tradespeople (two publicans, a wine merchant, a grocer, an auctioneer and ironmonger among them) who qualified as ratepayers. As such it is very unlikely they would have entered any of the houses of the poorer classes in Yarmouth before this point in time.

[10] *Norfolk Chronicle* — 10 May 1845, p.4.

2

The Victims

*"In the midst of life we are in death,
May ye be prepared to meet your God
Time swept, by His o'erwhelming tide
My faithful partner from my side,
And you of your's deprived
Maybe as unexpectedly as me."*
Verse found on the gravestone of Harriot Bussey,
St John the Baptist churchyard, Lound, Suffolk

By Monday evening the townspeople of Yarmouth were mourning the greatest loss of life to befall the town in their history. Its people were used to dealing with death by drowning. The winter storms wrecked ships in the Yarmouth Roads[1] and often bodies were washed up on the beach sometimes in large numbers. In 1807 144 bodies had been washed up after a heavy gale[2]. However, this disaster, apart from its unexpected nature, hit at the heart of the town itself and involved both the young and the innocent.

To get a better understanding of the effect of the disaster on the townspeople of Yarmouth one needs to explore who the victims and their families were and gain a sense of their status in society. This provides us with an insight not only into social aspects of this small coastal town in the mid-nineteenth century, but also might help explain how the disaster was and is remembered up until the present day.

As might be expected, given the time of day the event took place, the majority of the victims were female and under the age of twenty.

Many lived in the Yarmouth Rows which were narrow streets running east to west throughout the boundaries of the town walls. Originally these had names, and, while names were colloquially used in the newspaper reports of the time, in 1801 a numbering system had been formally adopted, with 1 being the most northerly and 145 the furthest south. As *Harper's Magazine* in June

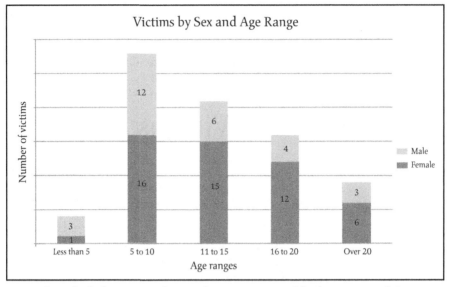

1882 described them, "Six feet is their average width, but some of them are scarcely more than three feet, and two persons cannot pass one another without contracting themselves and painfully sidling in opposite directions." Another

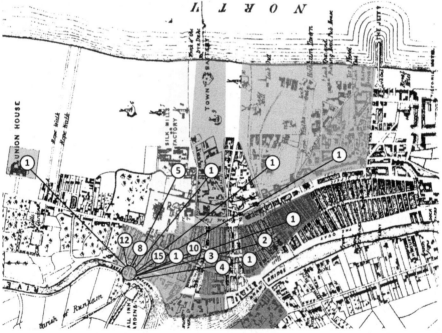

A map showing the number of victims by location using the 1841 census enumeration districts and an 1842 map of Yarmouth

description, supposedly written by Charles Dickens in Household Words Vol VII p.163, provides an insight into the paved Rows, which were to be found nearer the centre of the town. These differed from the narrower cobbled rows at the north end. "A few of the Rows are well paved throughout with flagstones ... If you want a stout pair of hob-nail shoes, or a scientifically oiled dreadnought, or a dozen of bloaters, or a quadrant or a compass, a bunch of turnips, the best in the world, or a woollen comforter and night cap for one end of your person, and worsted overall stockings for the other, or a plate of cold boiled leg of pork stuffed with parsley, or a ready-made waistcoat, with blazing pattern and bright glass buttons — with any of these you can soon be accommodated in one or other of the Paved Rows".

The 1841 census, in the main, used the numbering system and therefore it is possible to plot the areas affected using the descriptions of each enumeration district found in the census. By imposing the number of deaths onto a contemporary map of Yarmouth[3] we can see that the majority lived within close proximity of the bridge.

An analysis of the occupations of the heads of household confirms the view that the majority of the families affected came from poorer sectors of society[4].

This is borne out by the number that took up the offer from the bridge proprietor, Charles Cory, to pay for the victims' funerals. Of those that died, Cory paid for sixty-six to have a "decent internment"[5] leaving none to suffer the disgrace of a pauper's funeral. The number of victims who were buried in this way means that there is no physical evidence of their resting place. The cost of a child's funeral in 1850, according to the *Norfolk News*[6], was between £2 and £4. With many workingmen earning 20 to 25 shillings a week and sometimes a lot less[7], a child's funeral could plunge a family into serious debt. It would seem too that by the early 20th century, little had changed. When Maud Pember Reeves and Charlotte Watson did their influential survey of infant mortality in London between 1909 and 1913, called "Around About a Pound a Week"[8], they state that the cheapest a 6 month old child could be buried for was thirty shillings, and the older, and therefore bigger, the child, the greater the cost. It can be argued that the families selected for this survey were similar to many of the families in Yarmouth some sixty-five years earlier. They were not the poorest but could be described as the "respectable poor", with the head of the household in relatively stable employment, earning "about a pound a week". They would normally have subscribed to a burial club at a penny a week, taken out when the child was born with a small amount being claimable after thirteen weeks, or, if this could not be afforded, money would be borrowed from neighbours. Reeves', in her book describes a £2 funeral:

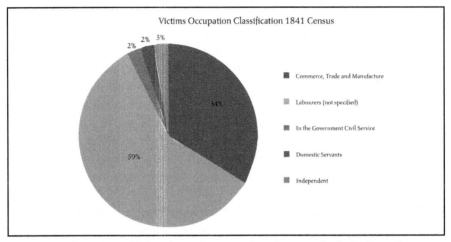

The three year old daughter of an out of work carter died of tuberculosis. The father, whose policies had lapsed, borrowed the £2 5s. necessary to bury the child. The mother was four months paying off the debt by reducing the food of herself and the five other children. The funeral cortège consisted of one vehicle in which the little coffin went under the driver's seat. The parents and a neighbour sat in the back of the vehicle. They saw the child buried in a common grave with 12 other coffins of all sizes, "We 'ad to keep a sharp eye out for our Ede," they said, "she were so little she were almost 'id."[9]

Where there is remaining physical evidence of the burial of victims from the disaster, it comes from those families who could either afford, or had insurance to enable, more than the basic funeral offered by Cory.

In the Dissenter's burial ground in Great Yarmouth seven of the victims were marked with stones, Jane Cole (aged 16), Hannah Field (aged 11), Alice Gotts (aged 51), Alice Gotts (aged 9), Harriet Little (aged 13), Matilda Livingstone (aged 6) and Joseph Livingstone (aged 7)[10].

Jane Cole was with Martha Field and her younger sister Hannah that afternoon. While Martha was saved, her friend and sister both drowned. William Field, her father, identified Hannah's body. He worked at the silk factory together with Maria Vincent, another victim of the disaster. She was employed as a silk throwster. Grout Baylis, a major employer in the town since 1806[11], had silk throwing and winding factories in Norwich, Yarmouth, Bungay and Mildenhall employing some 3500 workers in the region[12]. The majority of thread produced in these factories was then dyed and woven into mourning crêpe in Norwich or London.

An article entitled, "A Day at a Derby Silk Mill" published in 1843[13] gives a good idea of how William, perhaps as a foreman, and Maria as a throwster would have worked.

There does not seem to be any very definite distinction, among silk-throwsters, between the terms spinning, twisting, and throwing; or at least, the difference existing is not such as can be understood by general readers. ... Be it a "twisting" or a "spinning" machine, however, the action is both simple and beautiful. The floor or story in which these machines are congregated exhibits them ranged one behind another in two rows; and the eye is at once struck with the thousands of little spindles and bobbins which are whirling round at a very rapid rate, some yielding the silk which is to be twisted before reaching the others.

There is, to every machine, a set of bobbins whose axes are horizontal, and another set whose axes are vertical, and the twisting takes place while the silken thread is passing from the former to the latter. The vertical bobbins do not revolve, but they are placed upon steel spindles which pass through their centres; and these spindles, together with a kind of loop or eye attached to one end, revolve rapidly. The silken thread being passed from the horizontal bobbin through the eye or loop, and fastened to the stationary vertical bobbin, and motion being given to the apparatus, the thread becomes wound on the vertical bobbin by the rotation of the little loop apparatus, called the "flyer", round this bobbin; and a twist is at the same time imparted to the thread.

We have said nothing of the comparative velocity with which the two parts of the apparatus revolve; but it will be seen that a change in this relation produces a curious effect. If, while the bobbin maintains a uniform rate of movement, the flyer rotates more rapidly, the hardness of twist is increased, or there are more spiral turns in a given length of thread. If, on the other hand, the velocity of the flyer decreases while that of the bobbin remains uniform, or that of the flyer remains uniform while that of the bobbin increases, the twist becomes slackened, or there are fewer turns in a given length. The silk-throwster can therefore give any degree of hardness or closeness to the twist, by varying

A silk throwster.
The Penny Magazine, vol xii
no. 711. 1843. Pages 161 to 168.
(Supplement)

the relative velocities of the two moving parts.

However complex the twisting-machine may be seen at first sight, it is but a repetition of similar parts, each of which acts in the manner just noticed. All the horizontal bobbins are made to rotate by one piece of mechanism, while all the spindles owe their motion to another. The foreman or superintendent of the department regulates the relative velocities which the two movements shall bear to each other, according to the hardness of the twist to be given to the thread; but, when this is adjusted, women and girls attend the machines, replacing the lower bobbins when emptied, and the upper ones when filled, and also joining the ends of broken threads.

Alice Gotts and her daughter, also named Alice, had not lived in the town long having moved from Heigham, Norwich, some time between 1841 and 1844. William, her husband worked as an Excise officer, which meant he was moved regularly around the country. William's place of work was situated in a building on the north-west corner of Row 56 (Excise Office Row) while the family lived in the Conge. His work would have been concerned with checking and, if necessary, seizing goods that did not meet regulations, or items attempting to avoid tax via a thriving black market. Examples of the infringements in the town were regularly published in the local press:

> EXCISE OFFICE SEIZURES — Mr Greathead an officer of the excise, lately seized various quantities of spirits from different individuals, which were below proof. From the Feathers, he has taken two casks, containing about ten gallons, from the King's Head, a gallon of rum, besides a portion from the tap belonging to the same inn, and also a quantity from Messrs. Marsh and Barnes porter, who was in the act of taking it to Mr Page's, without a permit, which was to have been sent up in the morning. The latter may possibly be a matter for further inquiry, the rest are merely forfeited. A short time since a similar seizure was made from Mr. Thornton, which, we understand, the commissioners have ordered to be returned.[14]

Five months after the disaster (2nd October 1845), William married Elizabeth Starling. The couple did not stay long in Yarmouth and the 1851 Census finds them at Stowmarket, Suffolk. On 3rd December 1859 William died in the village of Bramford just outside Ipswich.

Outside Great Yarmouth two headstones can still be seen today that mark the graves of victims. In the small village of Lound, between Gorleston and Blundeston, can be found the grave of Harriot Bussey who lived with her husband George, a shoemaker, in Row 8 (called Ferry Boat Row as its west end was immediately opposite the ferry that the bridge replaced). Harriot would have had the shortest of journeys to make that day. Mary Ann Arnold testified she had seen her a little after five o'clock just off the bridge footpath standing by the inner palisades. Her husband George had identified her body and had touchingly made a chest to hold her remains, the material expense of which was defrayed as he belonged to a "box club"[15]. At the inquest he stated that he earned "from 5s. to 8s. weekly"[16].

The second grave can be found in St. Margaret's churchyard in Lowestoft. Reeder Thurston Balls was the son of John and Sarah Balls; a sixteen year old clerk at the Stamp Office[17] and described as a "youth of respectable connections"[18]. In 1845 he was living in Bath Place, but in 1841 was with his mother and older siblings Charles and Sarah in Jetty Road where his mother ran a lodging house.[19] He is buried with his grandparents William and Elizabeth Thurston.

However, the vast majority of victims were buried in St Nicholas churchyard Great Yarmouth in unmarked graves. Buried in batches, the most being twenty-seven in one afternoon:

> A dumb peal was rung in the morning, and immediately preceding this unusually solemn occasion; the whole of the service appointed by the Church was read over each body; the worthy minister being assisted by three or four clerical friends. The residents of the place and in the neighbourhood describe

The gravestone of Harriot Bussey, Lound churchyard, 2015

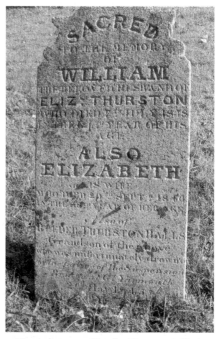

The gravestone of Reeder Thurston Balls, St Margaret's churchyard, Lowestoft, 2013.

the continual passing of bodies to the church, taken in connection with the melancholy event which was the cause of death, as one of the most distressing ever witnessed.[20]

Funerals at St Nicholas church as pictured in the Illustrated London News

Although a number of headstones in the graveyard have been lost[21], one has survived to mark the event.

The headstone of George H. J. Beloe must have been an expensive affair and is unique among the event artefacts that have survived. So much so it received Grade II listing in February 1998[22] and perhaps why it is worth investigating the family that commemorated their child further.

It depicts the fall of the bridge in relief based on illustrations produced at the time with the addition of the "eye of God" appearing in a burst of light through the clouds watching the collapse. The inscription reads:

> "SACRED
> TO THE MEMORY OF
> GEORGE H.J. BELOE
> THE BELOVED SON OF
> LOUISA BELOE
> *who was unfortunately drowned*
> *by the fall*

THE VICTIMS

of the suspension bridge
THE 2ND OF MAY 1845
AGED 9 YEARS.
Farewell dear boy no more I press
Thy form of light and loveliness
And those who gazed on thy sweet face
Knew it to be an angles[23] dwelling place
And if that realm where thou are now
Be filled with beings such as thou
From sin set free and sorrow freed
Then Heaven must be Heaven indeed."

George Henry John Beloe was the illegitimate son of Louisa Beloe, a straw hat maker[24] who lived with her mother, Margaret, and siblings Rachel, Henry, Clarissa Henrietta, and Matilda in Fuller's Hill.[25] It was Margaret that identified George at the inquest and gave evidence that he was on the footpath midway across the bridge. His mother, Louisa with one other child was reported as saved.

The top portion of the gravestone of George H. J. Beloe, St Nicholas churchyard, Great Yarmouth, 2013. The worn image is of the collapse of the bridge.

The family had probably moved from Norwich, where they lived in Surrey Street, to Yarmouth after the death of Margaret's husband John in 1836. Margaret (nee Mihigan) was born in Ireland in 1774. She married John Beloe (born October 1777) of Norwich in Cork in 1803[26]. In the 1812 poll book he is described as a shawl weaver; 1818, weaver, and by 1822, a Bombazine and Crêpe Manufacturer in St Augustine's. Finally in 1830 he is described as a "gent". He died on 7th December 1836 and was buried in St Augustine's, Norwich.

The Beloe family were therefore steeped in the Norwich textile industry. John had originally been a traditional "Norwich Shawl" weaver. These highly colourful shawls were the height of women's fashion in the mid-nineteenth century and commanded high prices. However, in his early 40s, he had moved into the "manufacture" of bombazine and crêpe. Bombazine is the material with a silk warp and worsted weft, usually dyed black, which in Victorian times became the chosen material for mourning. Black or white silk crêpe ribbons were used with the mourning garment depending on the relationship and age of the deceased or wearer. Norwich was the main city of manufacture for both types of cloth in the 19th century. Even by the 1820s, Norwich resisted any thought of organising its looms into "mills" like in Yorkshire. Instead, "Manufacturers" acted by either purchasing yarn, or in some cases, setting up spinning and throwing "factories", and retained a force of piece-work weavers working from their own homes. From 1812 the growth of power looms in Yorkshire put increasing pressure on the Norwich industry and it was failing to compete in the heavier woollen fabrics such as worsted. For a few decades, the Norwich industry remained viable as, "the lighter and relatively more costly mixed fabrics of silk and wool had been largely produced. Here the inherited skill of the Norwich weaver and the comparatively high value and light bulk of his product helped him in competition against superior organisation and machinery"[27].

There were four Beloes named as Bombazine and Crêpe "manufacturers" in Norwich in Pigot's directory of 1822. Arthur, John, William and Beloe and Co. It is very likely they were related to one another. The pressures of this period meant that the manufacturers and militant weavers, both of whom were highly organised into representative committees, often clashed when economic change confronted them. For example, on Friday 5th July 1822, the manufacturers' committee, which determined the piece-work rate, decided to reduce the payments made to weavers. A printed list of those manufacturers who had voted in the change was distributed and this led to uproar in the city. The journeymen weavers called a meeting of their number who were householders and therefore able to exert political pressure as voters[28]. 1100 of them met on the Monday and signed a petition for the manufacturers to reconsider. Alderman Robberds, the Deputy

Chairman of the manufacturers' committee, acceded to the request and called another meeting to be held on the Tuesday at the Guildhall. Further pressure was exerted on Monday afternoon as a huge number assembled on Mousehold Heath and drew up a second petition that representatives of the weavers be allowed to put their case to the manufacturers the next day. The *Norfolk Chronicle* reported the scenes of intimidation that followed:

> At the hour appointed for the meeting on Tuesday, a vast concourse of persons, of both sexes and all ages, had collected in the Market-place; and as gentlemen went up to the Hall, they were hooted or cheered by the populace just as they were considered friendly or adverse to the reduction of wages. On Mr. Arthur Beloe's making his appearance, he was received with loud acclamations. And, previous to entering the Hall, Mr. B addressed the multitude. He said that he considered the measure ill-advised. ... — "For my own part; (added Mr. Beloe,) I do not see the necessity for it; and as for the assertion respecting 400 pieces a week being sent from Yorkshire to this city to be dyed, I believe it to be a palpable falsehood. But be that as it may, I do not feel afraid of competing with the Yorkshireman, though he may be paying a penny or two per dozen less than we are. The old prices however shall be kept up by me." — This speech was followed by repeated shouts of applause and clapping of hands on the part of those to whom it was addressed.[29]

The meeting proceeded with all the manufacturers in attendance. Some argued that goods were being made in Yorkshire in such vast quantities and at such lower rates of pay that payments had to be reduced. Twelve journeymen weavers then addressed the meeting. Their statements focused on the increase in rents and coal arguing that it was impossible for them to live on a reduced rate. In addition, reduced rates would lead to inferior weaving and their goods would be brought into disrepute. The *Chronicle* continued:

> Up to this period of the business, nothing had happened of a seriously unpleasant nature: and the general deportment of the people was marked by a few aberrations from a peaceable course, as could perhaps be expected, in so many hundreds of them congregated together under circumstances of peculiar interest to their feelings. ... But in the course of the afternoon, an act of outrage was committed, not less serious in its consequences to the unfortunate subject of it, than afflicting to those who witnessed the scene, and disgraceful to the perpetrators. About half past four o'clock in the afternoon, as Mr William Bossley, a reputable manufacturer was proceeding up the steps of the Hall, in order to ascertain how the business

was going on, he was pushed violently down the stair-case by some of the persons with whom the porch was crowded; and in endeavouring to make his way into the Market-place, was beaten and kicked in a most dreadful manner. It was with the utmost difficulty that, by the spirited efforts of some humane individuals, Mr Bossley could be rescued from his infuriated assailants, so as to effect his escape into the shop of Mr. Woodhouse, in the Cockey-lane, who had some of his windows broken by the populace, for giving shelter to the almost exhausted object of vengeance. — Mr. B's hat, shoes and coat, which had been torn off, were afterwards rent to pieces and thrown up in the air in triumph.

In consequence of the turbulent and mischievous spirit which now manifested itself among the multitude, it was deemed expedient to apply to the Officer commanding the detachment of the 7th Dragoon Guards, stationed at our Barracks; that the military might be ready to act, when called upon by the Magistracy, in the aid of civil power, for the preservation of peace. — In the meanwhile, and shortly after the attack made upon Mr. Bossley, the conference between the masters and workmen had terminated. And it was announced from the balcony of the Guildhall, by persons who appeared to have just quitted the chamber where the discussion was held, that the manufacturers had consented to return to the Old Prices. — This declaration, as might be anticipated was received by the assemblage below, amounting at this moment to some thousands, with thundering shouts of joy and exultation, with waving of hats, and clapping of hands, which continued incessantly for many minutes. — But in stating the purport of what was thus publically communicated as a result of the conference, we beg to be understood as in no wise pledging ourselves for its correctness.
... We have, however, since been told that the master manufacturers perceiving the irritated state of mind under which the populace appeared to act, a majority of the meeting decided that it would be most prudent not to press the reduction of wages at this moment, but to leave the journeymen to be themselves convinced of the necessity of the measure.

... the Market-place and avenues to the Hall continued to be thronged with people. On Mr. Arthur Beloe's passing through, he was loudly cheered by them, and accompanied in a triumphant manner by a numerous body of weavers to his newly-erected manufactory on Orford Hill. A procession afterwards set out from the same place, with a band of music and flags at their head, and elevating a number of havels and sleas attached to poles, on which were fastened placards with the words "God bless Arthur Beloe" written thereon. Small specimens of the various kinds of articles manufactured by the weavers of Norwich were also carried aloft in procession, which paraded the principal streets till a late hour. A quantity of beer was given away to the populace.

...

No further disturbance of material importance occurred. The next day (Wednesday) was spent by many of the weavers in festivity and rejoicing, in various parts of our extensive and populous city.

By 1827, the manufacturers had begun to divide, some reducing payments while others, who perhaps had come from a weaving background, retaining higher payments. In February, a meeting of 4000 "operatives" met in Ranleigh Gardens to condemn those manufacturers who had reduced the wages of crêpe weavers without consulting all the manufacturers. At the meeting, John Beloe spoke together with Thomas Watts and Mr. Breeze.

Mr. BREEZE addressed the meeting at considerable length, ... he knew the price of poor man's labour and he knew its worth; he had been a journeyman himself and could speak to it that they were not a bit too well paid for their work.

Mr. THOMAS WATTS, a manufacturer, dwelt upon the impropriety of the meeting of the manufacturers by whom the reduction was determined upon; he said it was unneighbourly towards those who had not consented to it; it was a measure proceeding from persons who had been born with silver spoons in their mouths, and therefore did not know what was to be fed with anything else, and could not sufficiently appreciate the value of a poor man's labour.

MR. JOHN BELOE being called upon said that he coincided in the sentiments of the manufacturers who had already addressed the meeting.[30]

Despite his popularity with the weavers back in 1822, in 1829 Arthur Beloe was declared bankrupt and by 1830 John had retired, probably pleased that he had made enough money to provide an annuity for his wife.

We do not know why Margaret moved to Great Yarmouth following his death. Perhaps she realised that the economic prospects for her children in St Augustine's, Norwich were declining. George was born to Louisa soon afterwards, and there is more than a suggestion that her daughters may have been wayward. Despite Louisa being described as Mrs. Louisa Beloe in a list of those saved composed by one newspaper[31], it is clear she was unmarried and remained so[32]. Likewise, the baptism record for Great Yarmouth shows that a George Henry Beloe was baptised on 31 July 1846, the child of her sister Rachel, also unmarried. In the 1851 census return, there are two Beloe children in the Fuller's Hill household,

Henry J (aged 6, born in Norwich) and George (aged 4, born in Great Yarmouth). It is most probable that Henry J is the other child, saved with Louisa in that same newspaper list. As for the rest of the family, Louisa's younger brother, Henry, who was employed as a compositor in Yarmouth in 1841, married Sophia Bennett in Hoxton, Middlesex in December 1846. By 1851, Henry is living in Old Street, St Luke's, Finsbury, Middlesex with Sophia and 3 children (Henry William aged 4, John Arthur aged 2 and Edward Alfred aged 4 months). The first two are born in Hoxton New Town, Middlesex and the third in St Luke's, Middlesex. In 1861 the family were living at 33 Eaton Place, Hackney and had five children. His last entry is in the 1901 census where he is living at the age of 82 with his daughter at Frances Villa, Shelborne Road in Tottenham.

Clarissa, Louisa's younger sister also moved to London but is less fortunate and appears to have gone into domestic service before finding herself an inmate in the Holborn Union Workhouse in 1901.

Clearly a great deal was invested in George's gravestone. Whatever their fortunes following George's untimely death, their legacy was to leave, not only a very personal memorial to their loved one, but a wider reminder of this dreadful event for future generations.

[1] The Yarmouth Roads is a stretch of (usually) safe water which lies just off the coast slightly inshore of Scroby Sands. Ships would anchor in the Roads awaiting better weather. However, it could become dangerous in stormy weather with ships driven onto the sands. Turner's painting "Yarmouth Roads" in the Tate Gallery painted around 1840 illustrated the danger when a storm arose — http://www.tate.org.uk/art/artworks/turner-yarmouth-roads-tw0796 (last accessed 25 February 2015)

[2] Finch-Crisp, W. (1884), *Chronological Retrospect of The History of Yarmouth and Neighbourhood from A.D. 46 to 1884* available at: http://www.gutenberg.org/files/41618/41618-h/41618-h.htm (last accessed 3 April 2015)

[3] *Manning's Plan of Yarmouth, Gorleston and Southtown* (1842).

[4] The 1841 Census Commissioners were conscious of the need for a classification of occupations but were "fully alive to the difficulty of adopting any mode of scientific classification which would give general satisfaction". They did, however, devise the following broad classification of persons returned as: i. In Commerce, Trade and Manufacture; ii. In Agriculture (Farmers and Graziers; Agricultural Labourers; Gardeners, Nurserymen and Florists); iii. Labourers (a miscellaneous group but included all those whose employment "is not otherwise specified"); iv. In the Army and Navy; v. In the professions (Church; Law; Medicine); vi. In pursuits followed by other educated persons; vii. In the Government Civil Service; viii. Parochial, Town and Church Offices; ix. Domestic Servants; x. Independent; xi. Almspeople, Pensioners, Paupers, Lunatics and Prisoners. (In the chart shown, if no occupation can be found for the victim the occupation of the Head of Household has been used. It should also be noted that of the 47 counted in the "Labourers" class, 34 have no specified occupation or their family cannot be found in the census.).

[5] *Norfolk Chronicle* — 24 May 1845 p.3. — Cory asked Ben Dowson to enquire into those

that required assistance with burial costs. In a letter to the *Norfolk Chronicle*, dated 20 May he writes the following:

"To the Editor of the *Norfolk Chronicle*. Great Yarmouth, May 20th, 1845. Sir,—... I am authorised by the afflicted to return their grateful thanks ... to Mr. Charles Cory, the benevolent provider of the funds, which have been very considerable. The item of funerals shew, out of eighty sufferers, sixty-six have been provided with the expences of decent interment; ... I trust you will do me the favour to insert this in the *Norfolk Chronicle*, and you will oblige

Your's very obediently, BEN. DOWSON.", see also note 4 in chapter 4.

6. *Norfolk News* — 15 June 1850, p.2.
7. *Norfolk Chronicle* — 10 May 1845, p.4. – "George Bussey sworn — Was a shoemaker, in Ferry-boat row; had identified the body of his wife. Witness had procured a coffin for his wife, with the assistance of a club. He earned only from 5s. to 8s. weekly."
8. Reeves, P. (1914), *Around about a Pound a Week* available at https://archive.org/details/roundaboutpoundw00reevrich (last accessed 20 February 2015).
9. Ibid. p.70
10. See Chapter 1 for more information on the Livingstone family. Ages given are those on the gravestones.
11. Grout's first factory was established in 1806 on Northgate Street on the site of Lee's brewery. In 1815 it moved to the old barracks on St Nicholas Road where three years later a new five storey purpose-built mill was built. This building was damaged by fire in 1823 and more seriously in 1832 requiring a re-build. The factory used the latest technology to provide light having its own gasworks on site (established in 1829) which enabled day and night working. Source: Time and Tide Museum, Great Yarmouth. http://www.ourgreatyarmouth.org.uk/page_id__343_path__0p3p85p.aspx (last accessed 21 February 2015).
12. Coleman, D.C., (1992). *Myth, History and the Industrial Revolution* p.100.
13. *The Penny Magazine*, 1843 vol XII, No. 711. p.161-168. (Supplement) transcript available at: http://www.bednallarchive.info/misc/derbysilkmill_1.pdf (last accessed 20 February 2015).
14. *Norfolk News* — 3 January 1846, p.3.
15. *Norfolk News* — 10 May 1845, p.3.
16. Ibid., 7.
17. The Stamp Office would have collected stamp duty tax on all printed or legal documents where it was due. During the 18th and early 19th centuries, stamp duties were extended to cover newspapers, pamphlets, lottery tickets, apprentices' indentures, advertisements, playing cards, dice, hats, gloves, patent medicines, perfumes, insurance policies, gold and silver plate, hair powder and armorial bearings.
18. Ibid., 15.
19. *1841 England Census* for Great Yarmouth, HO 107/794/1.
20. *Illustrated London News* — 17 May 1845, p.316.
21. At least three gravestones did exist at the end of the 19th century that stated that death was due to the fall of the suspension bridge – Sarah Utting (aged 18), Mary Ann King (aged 11) and sisters, Mary Ann (aged 19) and Lydia Roberts (aged 11).

22 Listing available at http://list.english-heritage.org.uk/resultsingle.aspx?uid=1096818 (last accessed 21 February 2015).

23 The misspelling of "angels" on the headstone has been retained as it appears today. This error is likely to have occurred in the 1930s restoration of the stone (see note 32).

24 The 1841 census shows Louisa's occupation as "milliner", however the 1850 trade directory (1850 Hunt and Co· Directory of East Norfolk with parts of Suffolk p. 287) states she is a straw hat maker.

25 *1841 England Census* for Great Yarmouth, HO 107/793/7 6&7.

26 Irish Records Extraction Database - http://search.ancestry.co.uk/search/db.aspx?dbid=3876 (last accessed 23 February 2015).

27 Hawkins, C. B (1910) *Norwich: A Social Study* available at: http://www.forgottenbooks.com/readbook_text/Norwich_1000696490/17 (last accessed 23 February 2015).

28 Norwich was unusual in having a high proportion of its population able to vote. "… only Norwich and Nottingham had a franchise deep enough to allow radicals to make use of the electoral process." — Thompson, E. P. (1968). *The Making of the English Working Class* p. 513.

29 *Norfolk Chronicle* — 13 July 1822, p.2.

30 *Norfolk Chronicle* — 10 February 1827, p.2.

31 Ibid., 15.

32 *1851 England Census* for Great Yarmouth, HO 107/1806 3 gives Louisa and Rachel as unmarried and Henry J and George as sons of the Head of household, Margaret (a device often used for illegitimate children) whose age is given as 67, but more likely to have been 77. That said, the headstone in St Nicholas's churchyard says she died on 7th May 1831, aged 70. (Although the year date is a misinterpretation of 1851 by the stonemason when the stone was restored in the 1930s. A short article in the Yarmouth Mercury in 1931 describing the headstone transcribes the date as 1851.)

3

The Cause of the Collapse

> *"But sad catastrophe! terrific thought!*
> *The sight with life and all its sweets are bought!*
> *The faithless bridge spared not the concourse large*
> *But to the parting flood resigned its charge."*
> Consolatory Lines on the Downfal [sic] of the Suspension Bridge at Great Yarmouth On Friday, the 2nd Day of May, 1845; Addressed to the relatives and friends of those who were lost. By Mrs. Hart.

The Coroner, in his opening address to the jury, had given them the task of deciding not only how each of the unfortunate victims had come to die but also the more difficult duty of determining, "the cause of one of the most awful accidents that ever occurred in this part of the kingdom." The Coroner's powers, like today, were limited and he could not necessarily hold anyone to account. However, he would have been aware that the grieving people of Yarmouth would have sought reasons why this terrible event had happened to them. 19th century industrialisation, the control of "nature by man", brought different ideas about "cause". Man bore responsibility for change and should therefore ensure that scientific progress included learning from one's mistakes. The Sun summed up the prevailing feelings of the time when it stated:

> … we turn, therefore, from the irretrievable destruction of life to the precautions which the demands of future security so imperatively require to be adopted in testing the stability of all public works, structures, and edifices, upon or about which, masses of the people are likely at any time to be congregated. It would be strange, indeed, were this horrible casualty to be allowed to pass out of the public remembrance without furnishing a salutary as well as memorable lesson, in the first place, to those who undertake such constructions; in the second, to the authority, and principles, municipal or other, by whom they are employed. Is it possible to give more force and point however to such a lesson than are

communicated by some of the incidents connected with this deplorable accident?[1]

Although the coroner's jury had to be governed by practical reasons based on evidence; others did not. For example, the Rector of Great Yarmouth, Henry Mackenzie, had a great deal to say and based several sermons on the disaster. In his sermon on Whitsunday 1845[2], he clearly stated his opinions on the matter:

> The recent calamity — which I have before stated, I fully believe to be judgement upon your sins — that has befallen us, has already led your Ministers to speak to you in plain terms on the necessity of repentance and humiliation, faith and the amendment of life. ... Ignorance was undeniably the characteristics of those who were the sufferers. Ignorance and vanity (which in their combined action invariably produce Sin) are therefore the plagues shewn to be afflicting multitudes among us: and between these and our people I believe it to be our duty to stand, if we desire to do God's work in saving the yet living from being consumed.

SERMON,

PREACHED ON WHITSUNDAY, 1845:

Being one of a Series delivered after

THE FALL OF THE BRIDGE,

AT GREAT YARMOUTH;

BY THE

REV. HENRY MACKENZIE, M.A.,

INCUMBENT.

PUBLISHED BY DESIRE OF
THE COMMITTEE OF SUBSCRIBERS FOR THE RESTORATION OF THE PARISH CHURCH AND THE ERECTION OF A NATIONAL SCHOOL.

LONDON:
SMITH, ELDER, AND CO., 65, CORNHILL.

YARMOUTH:
CHARLES SLOMAN, KING-STREET.
GOOCH, MARKET-PLACE.

Frontispiece of sermon preached at Great Yarmouth by Rev Henry Mackenzie

While others blamed the outsider:

> ... never have we recorded so dire an event in a locality, proceeding from so trivial, so foolish a cause, as far as human actions can be considered the causes of unfathomable dispensations of Providence. The silly exhibition of a buffoon endeavouring to symbolise the very lowest degree of intellectual vagary, by the device of four geese drawing one — in the name and form so far as his mumming attire allowed a man — for the amusement of a gazing multitude — so childish an exhibition as that has been the immediate occasion of filling a whole town with grief and

hurrying into the presence of their maker above one hundred immortal souls![3]

The majority of the press, speculating on why the bridge had given way that afternoon, adopted a scientific rather than a religious or emotional response focussing on what changes had occurred recently, in particular, the addition of two walkways on either side of the main carriageway.

The suspension bridge and river, Great Yarmouth, Illustrated London News, May 10 1845

The Bridge is an elegant structure, suspended from two piers, and capable of standing a much larger number of persons than that we have named, but we understand every point of vision towards the spot where the geese were to be looked for, was densely crammed with men, women, and children, and even the chains and suspenders had many occupants. This is the Bridge which has been the cause of much litigation between the Yarmouth Railway Company and Mr Cory, its proprietor, and which has, since the arrangement with the parties, become the principle, if not the only medium of transit to and from the railway terminus. In order to accommodate the increased traffic, the proprietors have been induced to extend the bridge on each side of the chains, to the extent of four feet, for foot passengers, and the platform on the south side was the chief receptacle for the multitude who were on the bridge on this occasion; the north side was comparatively empty, consequently there was an extreme pressure

The fall of the suspension bridge, Great Yarmouth on Friday 2nd May, portrayed in the Illustrated London News, 10 May 1845

on the south — so much so, that a gentleman that passed over, noticed that the crown of the bridge, instead of maintaining its convex form, was completely flattened. He remarked the circumstance to a companion, but

at this moment all eyes were stretched to the utmost, and every ear listening with eagerness for the first announcement of the clown's appearance. This anxiety was brought to its highest pitch by the cry of "Here come the geese". The shout resounded from side to side, but amidst it was a shriek from the shores; the bridge was observed to give way; it lowered on one side;

the chains snapped asunder, one after another, in momentary succession, and in an instant the gaze of the thronging multitude was withdrawn from its object of worthless interest, and riveted on the half sunken bridge — suspended on one side by its unbroken chains — cleared of all its occupants — every one of whom was plunged into the stream, and over them the waters were flowing as though unconscious of the fearful tragedy which had momentarily occurred.[4]

During the identification of those that were on the bridge, the jurors were keen to explore from the witnesses whether the people on the bridge got any forewarning of the collapse, and whether, as a result, additional stresses were put on the bridge by the sudden movement of the crowd. Many of the witnesses at the inquest suggested that they had no warning of the collapse. While others said the collapse was not immediate and there was time in which an evacuation could have taken place.

A youth named J. B. Thorndike identified the body of Sarah Utting and deposed that he was on the Suspension Bridge when the accident took place. He stood in the carriage way and in about the centre of the bridge. He heard a cracking noise, and observed that one of the connecting links had broken, the ends of the severed link being about two inches apart. He did not apprehend any danger from what he saw. There was however, a sudden rush to the Yarmouth shore; but several persons laughed and then ran back to their places. To the best of his belief five or ten minutes elapsed between that time and the time when the bridge fell, when he, with others, was immersed in the water.[5]

At this point a link of the chain was shown in court with a break in it. Mr. Palmer, the foreman of the jury then spoke:

> By the Foreman: I had been on the bridge about five minutes before I heard the crack. It was crowded with people. There were a great many persons upon it, but they were not so thick as I have seen them. I should say they were about four deep. The bridge was not half full. Sufficient time elapsed between my first observing the link and the bridge falling for everyone upon it to have got off.[6]

> By a Juror — After the broken link had been observed, there was certainly sufficient time for the people to have escaped, before the falling of the bridge, had they tried.[7]

If this were true, the suspending chain had already been weakened by the weight of people on one side before any mass movement of the crowd towards the edge in their excitement to see the spectacle.

THE CAUSE OF THE COLLAPSE

On Tuesday morning the jury had completed taking identification evidence from witnesses and in the afternoon returned to the question of the state of the bridge. The jurors felt totally unqualified to consider the cause of the failure and Mr. Palmer suggested they should attempt to call a "scientific gentleman" as witness. The Coroner pointed out that he did not have the power "to procure the assistance of, nor to order any payment to any individual of scientific attainments to help them in arriving at a verdict"[8] and recommended that Mr Palmer, the foreman, who was on the borough council, ask whether that body would cover the expenses of an expert. Indeed Mr. Palmer had already drawn up a list of questions they might ask. These show that the focus of thinking was on the addition of the two footpaths on either side of the bridge the previous year.

1st. — By whose orders did you make the bridge wider?
2nd. — What was the extra width, and how many square feet were added?
3rd. — What was the weight of the cast and wrought iron with the wood placed extra on the bridge?
4th. — Did you offer or give up to Mr. Cory any mechanical ideas as to the proprietary of making the bridge wider, or explain the consequence of adding extra weight and breadth to act as leverage beyond the chains?
5th. — Did you know the weight the bridge was calculated to bear by the contractor, or inquire of Mr. Cory the terms of the specifications?
6th. — Did you examine the chains, bolts, bars, &c. before commencing the alterations, and report the same to Mr. Cory, as to their soundness and capability?
7th. — Did you calculate the extra weight you were going to add to the bridge before you commenced the alterations, and report same to Mr. Cory?
8th. — Did you suppose that if the bridge had been in its original state, would it have broken with the number of persons then on the bridge at the time of the accident?[9]

The borough council was to meet on Thursday and the jury passed a vote to request the following:

We, the undersigned, now acting as jurymen upon the inquisition of the bodies of the several persons who met their deaths by the falling of the suspension bridge over the river Bure, within this borough, beg to represent to the mayor, aldermen, and councillors of this borough, in council assembled, the absolute necessity that exists for employing some scientific engineer in order to come to a just consideration of the circumstances in which the bridge was placed, and to enable them to discharge the oaths they have taken.

Signed,   S. T. PALMER, Foreman.

Joseph Davey	Edward Garrod
James Emms	Thomas Stagg
Thomas Davey	John Orfeur
John Norman	John Fenn
William Spillings	Charles Barber
Mark Bowers	Charles Wolverton

The above were the only jurymen who signed the presentment at the time. [10"]

The council discussed the jury's proposal at their meeting. The advice of the Town Clerk, Samuel Tolver, their law official, was very influential in their decision not to meet their request. The clerk felt that, "... no useful result would follow sending for an engineer ... that even if a scientific person were sent for, and he pronounced there had been neglect in the building, that would have nothing to do with the death of the unfortunate parties.

> The Jury would not, as in cases of homicide, &c, send these cases for trial, nor could any parties be held to bail; and therefore no useful result could follow. The deaths of these parties had undoubtedly happened by the falling in of the bridge, and no engineer would be able to prove the contrary; and even if he entered into an investigation as to whether there had been a present defect in the bridge, or that its original construction was bad, he could not punish any parties. [11]

Mr. Cobb, made the observation that the Government had sent an engineer when seventeen lives had been lost at Ashton-under-Lyne when railway arches collapsed[12], and suggested that the jury should make a similar request to the Government. Ironically, the next day, the coroner's juries into the Ashton event were to issue a joint written verdict and they were clear that the cause was negligence:

> "We are of the opinion that this accident has occurred from the defective construction of the piers supporting the arches of the viaduct. The jury would specify the absence of a sufficient number of bond or through stones, and of the inferior quality of the mortar used. The jury feel compelled to add that, in their opinion, there has been neglect on the part

of those engaged, both in the construction and superintendence of the work." They also suggest that a copy of the evidence should be furnished to the Board of Trade, in order that the government inspector, who will have to survey the line before it is opened, may make a minute inspection of all of the works on the line.[13]

One Yarmouth councillor, "urged the Council to lend a helping hand to the jury on this subject", while another pointed out that, "The jury wished to go into a searching investigation of the case; for which the inhabitants would be indebted to them. He hoped the Town Council would not throw cold water upon them, but at least suggest something to prevail on the government to send down an eminent engineer."[14]

The jury received the news at their meeting on Thursday night, and unanimously resolved to petition Sir James Graham, The Home Secretary, to send down a government surveyor, and if they were refused they would start a subscription among themselves.

Clearly, the jury, as honourable men, felt as *The Times* on 10 May pointed out, "the enquiry should be full, fair and impartial" for the benefit of the townspeople and those that may or may not be implicated in potential negligence. However, the local newspapers, by concentrating on the addition of the two walkways, were stoking speculation:

> ...It is difficult to avoid the conclusion that the material, the design or the construction of this bridge, must have been faulty to a most extraordinary extent.

This was compounded by questions being asked of George Stephenson by the Parliamentary Committee considering the act to build the railway bridge over the Menai Straits.

> Mr. ESCOTT inquired if the Yarmouth suspension-bridge, the recent destruction of which has caused such a dreadful sacrifice of human life, was in principle the same as that which existed across the Menai Straits!
>
> Mr. Stevenson replied that the Yarmouth was altogether a badly constructed bridge, having only a single chain on each side; beside which, additions had been made to it for the purpose of accommodating an increased traffic, which had never been contemplated, and for which, in fact, it was evidently insufficient.[15]

We shall explore the relationship between the proprietors of the bridge and the Yarmouth and Norwich Railway Company in a subsequent chapter, but George Stephenson, was not entirely impartial as he was Chairman of the same railway company that had been taken to court by Charles Cory in the previous year. His opinion however was highly influential, and town councillors at their meeting referred to his remarks[16]. Without doubt, his comment would have been noticed in Yarmouth, something the *Norfolk Chronicle* was quick to point out[17].

We can get some idea of the tensions in the town with regard to discovering the cause of the disaster from *The Times*, who had sent their own correspondent to the town rather than rely on local reports. Perhaps he, as an impartial observer, was picking up the mood.

> The reply to this memorial [to Sir James Graham] is looked on for with great anxiety, and it is universally hoped that the prayer of it will be acceded to; for those few who objected to the town council paying for engineering evidence on account of the expense to the town, have not only no objections, but are anxious that such evidence should be had when it will cost them nothing.[18]

To the relief of all concerned, Sir James Graham agreed to send an engineer to inspect the evidence and produce a report for the jury. He appointed Mr. James Walker[19], who had only just relinquished the post of President of the Institute of Civil Engineers and was well known in Yarmouth, having been consulting engineer to the Pier and Haven Commissioners for some years. Mr. Walker, outlined four main conclusions to the coroner's jury[20]:

1. the immediate cause was a defect in the welding of the first bar that gave way in the chain on the south side of the bridge;
2. this was aggravated by the defective quality of the iron and workmanship. He concluded that these defects could have been noticed if a proper inspection had occurred when the bridge was constructed.
3. that the bridge had been widened without due regard to the strengthening that was needed.
4. that, in the original construction, no regard had been made for the weight being on one side.

Mr. Walker exonerated both Charles Cory, who had sanctioned the widening of the bridge, as well as the original architect, John Joseph Scoles:

> It has been stated, that the failure was occasioned by the leverage of the additional width given to the platform about a year ago. These additions certainly permitted a greater weight to be concentrated on one chain; but it is now generally admitted, there were only about three hundred persons on the roadway at the time of the accident, and supposing the platform had not been widened, they could all have stood on the same side, and

could have become concentrated on less than one-half the area of the platform. Had this been the case, the point of fracture would probably have been where it did occur; but if the chains had been sound and had been constructed of good iron, they would have borne a greater weight than could have been imposed upon them. As to the leverage from the increased width, this could not act prejudicially until after the instant of fracture.

The additions to the platform do, in some degree, exonerate the original designer of the bridge from responsibility, and if, as is stated, they were made without his concurrence, or without his revising the dimensions of the various parts, he must be entirely acquitted of blame.

The Greenwich memorial bust of James Walker.

It is singularly unfortunate, that the weakest links should have been placed at the point of the greatest tension. [21]

If Mr. Walker attached any blame for the accident, he felt that Mr Green, the appointed surveyor during the original construction, should have inspected the work more carefully. If this had been done, the accident would not have happened.

[1] *Norfolk News* — 10 May 1845, p.3.
[2] *Sermon Preached on Whitsunday, 1845 Being one of a series delivered after The Fall of the Bridge at Great Yarmouth*; by the Rev Henry Mackenzie M.A. Incumbent
[3] *Bury and Norwich Post*, 7 May 1845, p.3.
[4] Ibid., 1.
[5] *Illustrated London News* 10 May 1845, p297.
[6] Ibid., p.298.
[7] Ibid., 1.
[8] Ibid.
[9] Ibid.
[10] Ibid.
[11] Ibid.

12. *Yorkshire Gazette* — 26 April 1845, p.6. — on 18 April, during the construction of a branch line connecting Ashton-under-Lyne and Stalybridge on the Sheffield and Manchester Railway, nine of the twenty arches spanning the Huddersfield and Manchester canal and River Tame collapsed simultaneously killing seventeen of the workers who were levelling the line before the tracks were laid.

13. *Ibid.*, 1 p.4 (A full report of the experts' findings can be found in the *Manchester Times* — 03 May 1845, p.6).

14. *Norfolk Chronicle* — 10 May 1845, p.4.

15. *The Times* — 8 May 1845 — the bridge referred to was the road bridge built over the Menai by Thomas Telford and completed in 1826, three years before the one at Yarmouth.

16. *Ibid.*, 1. — "Mr. Richard Hammond was of the opinion that this important matter ought to receive the attention of the Council, particularly after the evidence of one of the most eminent engineers, Mr. George Stephenson, whom he (Mr. Hammond) had seen by the papers the day before. He had been before a Committee of the House of Commons, and gave it as his opinion that the bridge was not adequate to all the purposes for which it had been built.".

17. *Norfolk Chronicle* — 17 May 1845, p.3. — "The expression of the opinion of such a man as Mr. George Stephenson, is calculated to produce a very considerable effect on the public mind, and doubtless has produced that effect. But it will be evident to every impartial mind, that even such evidence must be valued only at what it is worth. —Not to notice the occasion on which Mr. Stephenson gave his opinion, there are many circumstances to be taken into the account. — If we could believe that Mr. S. really intended to say that the bridge itself, simply as a structure of that size and dimensions, was unsound or unsafe for traffic, how could he, as Chairman of the Yarmouth and Norwich Railway Company, allow it to remain, without indicting the proprietors or taking some measure to compel them to put it into a state of efficient repair?".

18. *The Times* — 12 May 1845.

19. James Walker succeeded Thomas Telford as President of the Institute of Engineers in 1834 and held that honour until January 1845. He was also chief engineer of Trinity House. He was very much considered, "a safe pair of hands" and "rubbished" Brunel's plans for a bridge over the River Severn in 1845 — see http://www.bbc.co.uk/news/uk-wales-19930031 (last accessed 2 March 2015).

20. *The Builder Magazine* — 31 May 1845, p.254.

21. Institute of Civil Engineers, *Minutes of Proceedings Vol 4 1845*, p.293 http://www.icevirtuallibrary.com/content/article/10.1680/imotp.1845.24457 (last accessed 2 March 2015).

4

Yarmouth Society

"between whom there is no intercourse and no sympathy; who are as ignorant of each other's habits, thoughts, and feelings, as if they were dwellers in different zones, or inhabitants of different planets...."
"You speak of —" said Egremont, hesitatingly.
"THE RICH AND THE POOR"
Sybil: Or, Two Nations, Book II, Chapter 5 — Benjamin Disraeli, pub. 1845

The disaster that befell the town of Yarmouth threw into stark contrast the almost unconscious divide between the working poor in the town and other sectors of society. In previous centuries it could be argued that the rich (or moderately wealthy) and poor lived side-by-side due to the confined area on which the town was built. There were all sectors of society living in the rows, albeit the wealthier merchant classes in the centre of the town:

> The residents in these rows are now principally mariners, fishermen, labourers, and the general poor; but formerly they were inhabited by a richer class, and many large and substantially-built houses still remain, which, although divided and in a mutilated state, give evidence that they were once the abodes of a wealthier grade of society. In fact some of the leading men of the town, down to a comparatively recent period, resided in the rows, which however were not then so much crowded by buildings, many of the houses having large gardens attached to them.[1]

Into these, often one room homes, the jury came to view the bodies of the victims:

> Never, perhaps, had any eighteen men such a painfully-distressing and truly melancholy task as devolved upon the jury. They assembled at three o'clock for the purpose of viewing the bodies of between 60 and 70 of their neighbours and fellow townsmen — they had to traverse the dirtiest and worst constructed rows in the town, some of which had as many as four or five bodies lying in each — in the great majority of cases they were conducted up staircases of barely sufficient dimensions to enable a

Row 1, circa 1870.

full-sized man to pass through, and on arriving at the summit were shown into the apartment where the body was lain forth, surrounded by relatives uttering the lamentations of distress and misery. In numerous other instances the parties were too poor to be able to occupy more than one room, and here were seen, in an incredibly small space, all huddled together, the living, the dying, and the dead; one unfortunate child recently shrouded, and with a parish coffin beside it in one corner; another child, suffering from the bruises occasioned by the fall and in extricating her from her perilous condition, in a second; and in the centre of the apartment, the mother suckling newly born twins! In one instance this was literally true. At a house in the North Entrance, where lay the body of Wm. Grimmer, a child eight years of age, the Jury found only one apartment for a husband, wife, and family of four children, and at the time of viewing the body, two children were lingering over a fire in a very infectious stage of small-pox. Two or three other of the houses visited were found to contain several cases of the above disease in its various stages.[2]

Those that formed the jury and the town council had very little need to visit the homes of the labouring poor in the rows, and, although they were aware that living conditions were bad, they had little first hand experience of the conditions. Within the town there were charitable organisations that tried to help the poor, one of which was the District Visiting Society[3]. This Yarmouth society did significant charitable work in the town, as can be seen by the report on the annual meeting in 1849:

> The annual meeting of the friends and supporters of this institution was held on Monday last, at the card-room in the Town-hall, the Mayor (P. Pullyn, Esq.) the chair, when there were between 60 and 70 ladies and gentlemen present.
> The old and valued Secretary, B. Dowson, Esq. read the report, which shewed that 1977 families had been relieved from the funds of the society to the amount of £415; and the amount of donations being about the

same last year, there remained balance in hand of £11 ...

The adoption of the report was then moved by the Rev. G. Hills, who in doing so was most anxious to express the deep interest which he took in the well-being of the society, which, as it was one for the moral and temporal improvement of the poorer classes of the town, ought to receive the support of all; and it was not only a benefit to the poor, but also a privilege for those engaged in carrying it out, for the more the two classes could be brought together, the better it would be for society generally. He begged to return his thanks to Mr. Dowson for his untiring exertions on behalf of the society, and he could but be glad that he was associated with him in his good work.[4]

It was the ladies from this society that visited the sufferers immediately following the disaster:

> To the Editor of the Norfolk Chronicle. Great Yarmouth, May 20th, 1845.
> Sir,—Directly after the melancholy accident at the Suspension Bridge, I was consulted, as Honorary Secretary of the District Visiting Society here, on the probability of the ladies attached to that institution being induced to devote their kind services as agents in administering to the wants of the survivors at the late sad catastrophe, as well as the relatives and friends of the sufferers, if funds were provided. I have the satisfaction of stating, immediately the application being suggested, the ladies came forward with great willingness, zeal, and alacrity; visiting the dwellings of all who were in need of comfort and relief, and I trust there has not been a case overlooked. In every instance that has come to my knowledge (and they are many), I am authorised by the afflicted to return their grateful thanks to each visitor, for her kind sympathy and attention ... nor has any applicant for injury or loss sustained, that has come to visitors or my knowledge, been turned aside without assistance, after careful investigation. The exigency for ladies' services being nearly finished, and the account about closed, I trust you will do me the favour to insert this in the *Norfolk Chronicle*, and you will oblige.
> Your's very obediently, BEN. DOWSON.[5]"

It is clear that the political, social and economic elite of the town, those men who made up the town council, magistrates, justices of the peace, and poor law commissioners, many of whom were actually the same people, knew only too well of the poverty that existed among the labouring poor, but could do little about it. The ratepayers were no different than others throughout the country to objecting to the costs that befell them and in some cases even refused to pay if they felt the finances had been poorly managed or were too much[6]. Twenty years later in 1864 there were still concerns about diseases such as smallpox and the unsanitary conditions in the northern rows:

> Among the reasons that have tended to increase the unfavourable opinion entertained respecting the healthiness of Yarmouth has been the fact that many apprentices belonging to smacks have been landed at this port suffering from small-pox. They have been taken to certain rows at the North-end … Most of the cases that have occurred in Yarmouth have been of a mild type; and as the disease is now on the decline, it is to be hoped that in the course of a short time it will altogether disappear from the poorer localities, to which it has been confined. It is in some degree to be regretted that more attention is not paid to cleanliness in the greater portion of the rows. To judge from their appearance they are not very frequently swept, while offal is often allowed to accumulate to an unpleasant extent. Additional sanitary measures are unquestionably desirable in the more densely populated parts of the borough, especially during the summer season.[7]

The two popular Victorian schools of thought, Utilitarianism and Political Economy[8], both played their part in the attitudes of the townspeople. However, economic and political theories were not the only things that shaped attitudes towards the poor. Religion also played a major part. The evangelical revival of the previous century was based on the teaching that conversion could rescue all but the most degraded in society. The clergy's task was to rouse people of all classes to the worship of God:

> The Evangelicals were by far the most active members of the established Church and exercised tremendous influence from the end of the eighteenth to the middle of the nineteenth century. It has been said that more than any other single factor the evangelical movement in the Church of England "transformed the whole character of English society and imparted to the Victorian Age that moral earnestness which was its distinguishing characteristic".[9]

It perhaps then is no surprise that Rev. Henry Mackenzie, incumbent at St Nicholas Church, was to use the disaster to promote this enthusiasm. In his sermon of 11 May he made clear that, "God has lately sent his destroying angel among them, for the vanities of this wicked world" and "it was the duty of those assembled to aid the work which he proceeded to indicate". [10]

I do not bring forward my suggestions for making this stand without long, thoughtful, and prayerful consideration. From the time that GOD sent me amongst you, I have been pondering on the means of increasing,

1. The accommodation for the poor in this House of God; and
2. Education for the children of the poor in the principles of Christ's Church.

First then, I have reason to state with confidence, that this Temple of the Lord may be so arranged as to accommodate, in more seemly a manner than we now exhibit, both as to convenience and beauty, an additional number of at least five hundred of our poorer brethren, as well as largely increasing the extent of pews or appropriate seats.

Next, I believe, through the kindly aid of patrons of the parish, we may be assisted, at comparatively trifling outlay, to the possession of an ancient and most suitable building for the purpose of establishing a National School for at least two hundred children, to be there reared in the nurture and admonition of the Lord.[11]

Reverend Mackenzie's estimate was that £5000 was needed to rearrange St Nicholas church and create an additional 500 seats for the poor of the parish, and to establish a National School[12] using the adjacent old Priory buildings, "which are now used as a *stable*, but which, at a moderate expense, can be converted to this purpose".[13]

By December 1845, he had secured £1250 from the Church Trustees, and raised another £1439 5s from subscriptions, one subscriber being Rt. Hon. William Gladstone MP who gave 2 guineas towards the establishment of the school.[14] In late August 1848, the church restoration was complete.

Rev Henry Mackenzie
Perlustration Of Great Yarmouth With Gorleston And Southtown, C.J. Palmer. 1872

On Thursday week, this noble parish church, after complete restoration and renovation, was re-opened for divine service. The solemnities of the day excited much interest in the town, and the morning and evening services drew large congregations, including a number of persons from the entire eastern district. The morning congregation comprised about 3,500 persons, and hundreds could not find seats.[15]

The establishment of the school took longer. However, in September 1852 The Bishop of Norwich opened three schools (boys, girls and infants) on the site of the old Priory. During his sermon he spoke as follows:

> He next adverted to the great impetus which had of late years been given to the education movement. From various motives a general awakening was apparent. Some had assisted it from mere charity — as they would contribute to a hospital — whilst others had endeavoured to develop some scheme or theory of their own as a remedy for all the social evils existing among us. We talk, said his lordship of the power of the press, but a still mightier agency is at work in the plain unpretending buildings, which we see every where grouped with churches and chapels in our villages, or retiring from notice amongst our stately streets and imposing edifices of our towns. We are rocking the cradle of a giant. But can we say that in all this, our interest is of a kind which needs not our Saviour's admonition. Can we say that we are now fully alive to the importance of the children of the community. I am afraid we can hardly say that. ... are we zealous for the general education of the children of the community in those religious principles which we can alone make that education a blessing? Let us not turn away from the contemplation of our duties and responsibilities to the poor and vicious, because there is humiliation and disgrace in it, and danger too — yes, danger. England has under Providence, I trust but little to apprehend from without. Her sources of danger of real danger are from within. There is throughout this country, and especially

St. Nicholas church, Yarmouth
Meall's pictorial guide to Great Yarmouth, 1854

in our larger and more populous towns, a mass of wretchedness, and ignorance, and demoralisation, which on our own account, if not for other considerations, we cannot contemplate without anxiety and alarm. We do not often see them. We do not even know them. They are not within the sphere of common sympathies. They are people away, out of sight. They lie between the surface of society, like the combustible materials hidden beneath the volcanic mountain. We are made aware of their condition, from time to time, by some startling movement which disturbs society; something that forbodes trouble to us[16].

Hall of the Benedictine priory (restored 1852)
History of Great Yarmouth, Henry Manship 1854

The fall of the suspension bridge some seven years before had indeed been a "startling movement" that had disturbed the society of Yarmouth, highlighting, if only for a short time, the conditions of the poor within their town. Although little might have been done to address the grinding poverty of the working poor, the foundation of the Priory schools was the most enduring result of the disaster. As the bishop finally remarked, "The three schools for which I plead include one for boys, one for girls, and one for infants, and are calculated to bestow the blessing of education on 550 children. They are not the sole schools for the education of the poor in this district, but I can affirm that they supply no more than is adequate, that is requisite, to make up for the deficiency of school instruction."[17] While new infrastructure to support commerce in the town could be regarded as inevitable, new institutions to support the education of the poor were not.

[1] Palmer, C.J. (1872) *Perlustration of Great Yarmouth with Gorleston and Southtown*, p.19 available at https://openlibrary.org/books/OL23649652M/The_perlustration_of_Great_Yarmouth_with_Gorleston_and_Southtown (last accessed 4 March 2015).

[2] *Illustrated London News* — 10 May 1845, p.298.

[3] Elizabeth Fry had established the first district visiting society in 1824, following a holiday

in Brighton where she was disturbed by the number of beggars on the streets of that town. Her scheme was to establish a team of volunteers who would go into the homes of the poor offering help and comfort. The idea spread quickly and soon towns up and down the country had their own visiting society.

4   *Norfolk News* — 4 August 1849, p.3.

5   *Norfolk Chronicle* — 24 May 1845, p.4. see also note 5 in chapter 2.

6   *Norfolk Chronicle* — 28 November 1840, p.3. — One of the protagonists in our story had done so in 1840. Charles Cory, the proprietor of the bridge in 1845, together with Robert Ferrier, appeared before the Justices of the Peace to answer to summons issued against them for non-payment of the poor-rates. Both admitted that they had not paid, as it gave them "an opportunity for publically noticing that by the accounts of the Clerk of the Guardians, the parish had occurred a debt of £2100 2s 2d prior to the 25th March 1840. … Mr. Cory contended that the rate was not good, insomuch as it was excessive, and the old rate was not collected up. … The Mayor asked My Cory if he refused to pay the rate. Mr Cory did refuse: a distress warrant was immediately issued, and a clock, a silver cup, and a hat-stand were taken from his premises. We understand that Mr. Cory intends to apply to have the goods sold by public auction.".

7   *Norfolk Chronicle* — 2 July 1864, p.6.

8   *Hard Times: fact and fancy* — article by Paul Schlicke — "Utilitarianism, the philosophy of Jeremy Bentham, taught that human nature was motivated by self-interest, and it was the duty of the state, through education and extension of the franchise, to support and encourage each individual to pursue his or her own interest. Political Economy, on the other hand, considered national prosperity dependent on unchanging economic laws, according to which the individual self-interest of employers promoted the general welfare. Raising the wages of workers, the followers of Adam Smith and David Ricardo taught, would cut into the profits of industrialists and thereby jeopardise national prosperity. The proper role for the state was one of laissez-faire, to enable the process to operate freely." - See more at: http://www.bl.uk/romantics-and-victorians/articles/fact-versus-fancy-in-hard-times#sthash.lGjrh45H.dpuf (last accessed 4 March 2015).

9   Young, A.F. and Ashton, E.T. (1965). *British social work in the 19th century*, p.28.

10 *Norfolk Chronicle* — 17 May 1845, p.3.

11 *Sermon Preached on Whitsunday, 1845 Being one of a series delivered after The Fall of the Bridge at Great Yarmouth*; by the Rev Henry Mackenzie M.A. Incumbent

12 The National Society for Promoting Religious Education created "National Schools" from 1811 adopting the monitorial system as its means of instruction, with its curriculum based on the teaching of the Church of England. The aim of the National Society was to establish a national school in every parish in England and Wales.

13 *Norfolk Chronicle* — 6 December 1845, p.3.

14 Ibid.

15 *Cambridge Independent Press* — 2 September 1848, p.1.

16 *Norfolk Chronicle* — 9 October 1852, p.2.

17 Ibid.

5

The Building of the Bridge

"On coming from the Bridge I picked up a bent sixpence in the Road which was looked upon as a lucky omen."
Narrative of the Commencement and completion of the Suspension Bridge over the River Bure at Great Yarmouth erected by Mr. Robert Cory jun. in 1828 and 1829, R. Cory, 1832

The tragic event on 2 May 1845 was in stark contrast to the opening of the suspension bridge some 16 years previously. In fact the story of the bridge, during its short life, perfectly illustrates the pressures within this small and ancient coastal town during the economic, social and political upheaval of the 1830s and 40s.

Yarmouth's economy by the mid to late 1820s was in danger of decline. While new industry, such as the Grout Baylis's silk factory was flourishing, the end of the Napoleonic Wars meant that the military presence in the town had diminished. The town's mainstay was its traditional herring and mackerel industry, which was unable to expand as quickly as some wished due to the town's poor transport infrastructure. Most of Yarmouth's products were transported by sea to London, or river to Norwich. By the 1820s Yarmouth was significantly behind others in improving its transport links to enable the economic growth of the town. However one man was willing to take a unilateral decision to invest in the future, and to consider his legacy by recording his actions for posterity.

Robert Cory Jun., a solicitor in the town, was a member of the town's political class[1]. His father had been Mayor in 1803 and he himself had held the same office in 1815. In the 1830s he began to

Robert Cory Jnr

write an unfinished history of the town. Among these papers was a full account of his endeavours to build a suspension bridge[2]. He starts his account by stating:

In 1810 I purchased an Estate at Runham opposite Yarmouth North Quay comprising a Public house and Gardens called Vauxhall of about 100 acres of land together with a ferry and tolls over the River Bure.

Long before I purchased the Estate I had entertained an idea that at no distant period a Bridge would be made across the River hereabouts and a Turnpike Road from Yarmouth to Halvergate and so five years after about 1823 I projected such a Road and went to a considerable expense for plans and estimates, gave notice of an application to Parliament and formed a committee of Gentlemen to carry it into effect, but it was thought that the Tolls would not meet the expense and the measure was abandoned.

Afterwards several Gentlemen and solicitors undertook the same plans knowing it must materially benefit my Estate, I rendered them every assistance, however they are proved abortive.

One Morning in August 1826 coming from Horsey, I observed several sticks set up at various distances with papers on them on the Caister Road about a quarter of a mile from the Gates and on enquiry I learnt that a project was on goal for making a turnpike & building a Bridge across the River near the Cinder boons, and that these sticks were the surveyors' marks. This project would have avoided my land and made my Tolls valueless. Without disclosing my intentions, I immediately gave notices for an application to Parliament for an Act to build a bridge at my ferry, and before it was generally known, I obtained an Act which received the Royal assent 20 May 1827.[3]

A view of Yarmouth on the Bure or the North River, 1826.

Cory deliberately inserted a clause in the act preventing anyone from establishing a bridge or ferry within one mile of this point so securing tolls on all goods or persons wishing to cross and use the new road that he also hoped to build.

By May 1827, Cory was ready to embark on his project and advertised for plans. It seems from the outset he was set on an iron suspension bridge.

> I had often seen Brighton pier, and had some inclination for a chain bridge, and in October 1827, I went to look at one at Witney which had been built by Mr. Gibson. Mr. Goddard accompanied me and took the measurements, and soon afterwards I went to Brighton and on my way looked at the Hammersmith one. I then saw a small one at Ryde in Sussex, and afterwards, hearing there was one erected at Leeds upon a different principle, using rods being suspended from an arch thrown over the River instead of from chains over pillars. In January 1828 I went and looked at it, but did not like it, and determined upon one suspended from chains.[4]

Image Brighton from the Chain Pier By N. Whitlock, 1829

By this time he had employed a local builder, Godfrey Goddard, to make plans and models for the bridge. It seems that Robert's eldest son, Isaac, was not impressed with these, saying that such a structure "would not stand an hour". He suggested that his father should call in a reputable architect, and recommended

John Joseph Scholes, a London man whom he had met in Italy. Goddard was therefore instructed to build a bridge designed by Scholes, and on 9 June 1828 work began. By 28 July the foundations were complete and Mr Cory laid the first stone, which had been hollowed out to receive a few coins of the realm and a parchment commemorating the event.

The building work was completed by February 1829 but Cory decided to wait until 23 April, being George IV's birthday, to open the bridge.

The opening was a grand affair with invitations sent to all principal inhabitants and corporation officers. Cory describes the grand procession from the Guildhall to the bridge and across, where a lunch for 200 guests on the green was waiting. With the town band playing some 400 to 500 children sang God Save the King. The day ended with a ball held by Cory. The town were so grateful that a "number of gentlemen" subscribed to the minting of a commemorative medal, one silver, for Cory himself, and fifty copper to mark this important occasion.

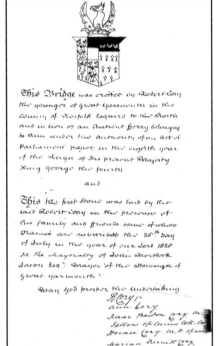

Parchment placed in the foundations of the suspension bridge, 1832

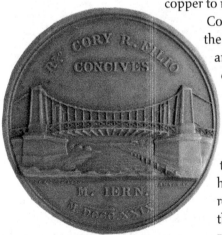

The reverse of the suspension bridge medal of 1829.

Cory's account shows an insight into both the man and the time. A local landholder and gentleman, he was entrepreneurial enough to consider capitalising on the road revolution that was transforming the country's industrial infrastructure (1830 was the peak of Turnpike Trust Acts passed). The only route at the time was around the marsh (this road had been turnpiked in 1769). A faster route to Norwich would not only benefit the town but also provide significant revenue concerning his rights over the Bure. Building the bridge was no small financial undertaking. His detailed accounts showed that the bridge cost £3285 10s, and at the time, with a further amount needed to be raised to build the turnpike road, this was a significant risk. Cory's financial risk paid off as, once the bridge was built, he attracted capital from other investors to complete the road[5] across the marshes to Acle:

> In the following year I set myself heavily to get a Turnpike Road from my Bridge to Acle which notwithstanding many opponents, I fully succeeded in "an Act for making and maintaining a Turnpike Road from the Bridge over the River Bure at Great Yarmouth to Acle (with certain Branches therefrom) all in the County of Norfolk" received the Royal Assent, the 3rd May 1830 & the Road was immediately after begun.[6]

One final perhaps significant event took place in 1837. Robert Cory decided to reposition the tollhouse, which was originally on the Yarmouth side of the river, to "the Runham side of the water about 100 yards from the foot of the bridge":[7]

> The overseers of Yarmouth having laid me heavily to the poor Rates & the inconvenience of pulling Horses up to the Toll on the descent of the Bridge being great, I determined to build a Tollhouse on the Runham side which I did. It was planned by Wilson John (who is with Mr Scoles the Architect as a pupil) & was begun in April 1837 and completed in Sept. following - To this I attached a piece of ground as a garden - John Fountain who I first appointed collector of the tolls died in March 1837 and I appd Charles Bailey who took possession 25 March & went into the new house, & began to collect tolls there 29th Sept.
> "The former Tollhouse I now let for £5 a year." [8]

This meant that in May 1845, when the townsfolk gathered on the bridge they would not have passed through the bridge's tollgate and no payment would necessarily have been demanded. Perhaps, if they had had to pay the halfpenny required of foot passengers to cross the bridge, some might have chosen a different position to view the scene.

One might wonder why Robert Cory Junior might have been so interested in acquiring the Runham estate and invest in infrastructure projects. Both he, and his father were solicitors in the town, as indeed were his sons. In examining the local newspapers for the period, their practice involved mostly selling and buying land and properties for other people, or handling the estates of bankrupts, rather than speculating in property themselves. It could be said that the bridge was somewhat of a vanity project to secure his legacy in the town. A fellow of the Society of Antiquaries, he had started writing *"Miscellanies relating to Gt. Yarmouth"*[9] which featured a whole chapter on the building of his suspension bridge, and found in

The suspension bridge, 1841.

his large personal library on his death. He was also interested in coins; perhaps this is why the thought of a minted coin to celebrate the opening of the bridge was deemed an appropriate gift by his contemporaries. Cory adorned the bridge with his crest and iron plates announcing that he had built it and certainly the placement of coins and a scroll in the foundations suggested he assumed it would stand for some time as a testament to his endeavours, if not ambitions for the town. The fact that on two previous occasions he had failed to attract investors, and was willing to take the risk on himself, suggests motives other than just pecuniary. On deciding to build the bridge he could have had no guarantee that the road, with much more lucrative potential, would have gathered other investors. However, without a bridge that project would never be realised. Perhaps Cory was shrewd enough to realise that the smaller investment necessary to build a bridge would remove the barrier for further investment by others. Critically though, from his description, it would seem that protecting his rights was the major motivation. He perhaps knew that it was those rights that were the most valuable asset in any future enterprise.

The year before Cory had been galvanised into considering building his bridge, the Stockton and Darlington Railway had opened using Locomotion No 1, the first passenger carrying movable engine, built by Robert Stephenson. In 1823 George Stephenson, his father, had built the Gaunless Bridge for the line, the first iron railway bridge[10]. While Stephenson was concerned with engineering strength, Cory was concerned with aesthetic design.

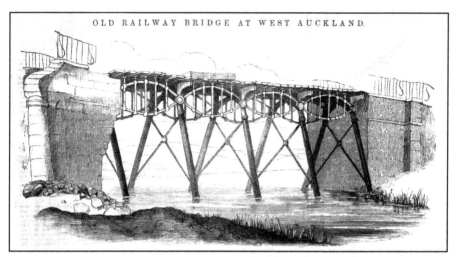

The old railway bridge at West Auckland, 1875

It is somewhat ironic that, despite the fact that 1830 saw the peak of Turnpike Acts being passed by Parliament, the future of transportation for Yarmouth and the rest of the country lay in iron railways. His bridge, whether it had collapsed or not, would have made way for something more appropriate on completion of negotiations between his son and the railway company the very month of the disaster and his legacy to the town would be very different from that he envisaged.

[1] Civic offices in Yarmouth were held by a small number of families. During the 1834 Royal Commission into the state of corporations that led to the Municipal Corporations Act of 1835, the relationship between members was explained. "The inquiry before Messrs. Hogg and Buckle on the 1st day related chiefly to the constitution of the body, and mode of electing the Mayor, in which some curious manoeuvres have been practised. On the 2nd day, Mr Alderman Barth made the following statement of the family groupes in the Corporation:— John Preston, Edmund Preston, and Isaac Preston are brothers; Mr Reynolds brother-in-law; Isaac Preston jun. son of Isaac Preston; all Aldermen. In the Commons, Mr. I.K. Preston, son of John Preston; E.H.L. Preston, son of Isaac Preston; and Frederick Preston, cousin of Isaac. The Major, J.D. Palmer, Esq. In the Common Council, G.D. Palmer his nephew; Ambrose Palmer, a nephew; and Charles John Palmer, his son. Robert Cory Esq.; Robert Cory jun. his son; Alderman. In the Common Council, S.B. Cory, also his son. William Palgrave, Alderman. In the Common Council, William B. Palgrave, his son; and John Ker, his son-in-law. Thomas Bateman, and George Bateman, Alderman. Charles Costerton, Alderman. In the Common Council, George Costerton, his brother, also related to the Mayor by marriage; and Samuel Costerton and Henry Costerton, brothers. Sir E.K. Lacon, Bart J.M. Lacon, his brother; Aldermen. In the Common Council, J.E. Lacon, son of Sir Edmund. Robert Fenn and Robert Bould Fenn, his son, Common Councilmen." — *The Bury and Norwich Post* — 19 February 1834, p.3.

[2] *Narrative of the Commencement and completion of the Suspension Bridge over the River Bure*

      at Great Yarmouth erected by Mr. Robert Cory jun. in 1828 and 1829, R. Cory and bound in "The Suspension Bridge" dated 1832. See Appendix.

3   Ibid.

4   Ibid. —The Royal Suspension Chain Pier was the first pier in Brighton and was built in 1823. It is the subject of paintings both by Constable and Turner.

5   *Norfolk Chronicle* — 23 April 1831, p.2. — "Perhaps it is not generally known, that the saving of distance from Acle to Yarmouth, by this new road, will be three miles and five furlongs, and a great advantage of this project is, that by means of branches, a large tract of country will be laid open to Yarmouth, which has hitherto been nearly excluded for a great part of the year, on account of the distance of roads by a very circuitous route. In addition to the satisfaction which the trustees feel at the progress of the works, the shareholders have the gratification of knowing, that they will receive ample interest for the principal money, invested so judiciously, and with so much public spirit in this useful undertaking."

6   Ibid., 1.

7   *Suspension Bridge* from *Miscellanies relating to Gt. Yarmouth by R. Cory with additional notes by J. Davey* dated 1846. See Appendix.

8   Ibid.

9   This hand written chapter included a note added by Joseph Davey dated 1846, one of the inquest jurors, "On the 2nd of May 1845 a number of persons having assembled on the bridge to witness a Clown from Cooke's Equestrian Circus swim in a tub drawn by geese, one of the suspending chains gave way & precipitated the mass of people into the water, by which accident 79 lives were lost. It is very remarkable that of this large number not one head of a family was removed so as to leave the survivors dependent on parish relief. The scene was the most distressing I ever witnessed. J.D."

10 Stephenson used a lenticular truss for his bridge. Lenticular trusses were patented in 1778 by William Douglas, although the Gaunless Bridge of 1823 was the first of the type. This design uses two curved girders in a lens shape, one above and one below. The unusual feature of the Gaunless Bridge was that the wooden deck is above the truss. This allowed Stephenson to use cast iron for the vertical members, rather than more expensive wrought iron. Since 1975, it has been on display in the car park of the National Railway Museum, York.

6

Road, Rail and Bridge

"You see, Tom, ... the world goes on at a smarter pace now than it did when I was a young fellow ... it's this steam, you see, that has made the difference; it drives on every wheel double pace and the wheel of fortune along with 'em."
Mr. Deane to Tom Tulliver
The Mill on the Floss — George Eliot, pub. 1860.

The bridge and turnpike road remained the main means of passage, apart from the river, between Yarmouth and Norwich for some fourteen years. However, in 1842 the Yarmouth and Norwich Railway Company was established with a proposal to build a line between Norwich and Yarmouth[1]. George Stephenson, the great railway engineer, was elected Chairman of the Company, and his son Robert appointed engineer.

At the Yarmouth end of the line, the Cory family's rights over the Bure had to be considered and consequently the railway station fell short of coming into the town, being built on the Runham side of the river at Vauxhall. The railway's Act of Parliament, passed on 18 June 1842, expressed the compromise reached between the company and the bridge owners. In order to protect their rights, the Corys had to agree to any future bridge being built over the Bure by the company.

However, as progress was made towards the building of the line, the Directors[2] of the railway company became discontented with their decision to allow a third party so much effective control over their actions. In 1844 they put an additional bill before Parliament[3], and, without consulting the Corys, included the wish to build another bridge over the Bure further south at Lime-Kiln wharf. The fish merchants supported this in particular, and wanted to ensure that their produce could be easily loaded into wagons on the quay and from the seafront jetty "without the obstruction which a toll would occasion" to the Vauxhall terminus. 191 of them petitioned the Committee of the House of Commons considering the 1844 bill to this effect[4]. The Mayor and one other Town councillor, Mr. R. Hammond, also signed this petition, but others had refused[5].

Therefore, when the bill was published many were surprised:

From a Correspondent — The Directors of the Yarmouth and Norwich Railway Company have printed their bill. It appears, that they require authority to make a bridge over the Bure or the Yare, and to make either of the lines laid down in their plans. In their references all the land and houses along the North Quay are included as numbered on the plans; and though they have scheduled only sufficient to enable them to place the bridge where they choose, which induces us to suppose that all they want would be as much space as would enable them to use the intended bridge without inconvenience, the bill empowers them to *make wharves*, &c., evidently bespeaking their intention, if possible, to carry out the scheme of sweeping the North Quay, and appropriating to themselves or friends a monopoly of trade of this port, at least as far as the inland traffic is concerned. But what surprises us more than anything else in the bill is the attempt to vest the whole of a piece of land banked in from Breydon in the Company, after a solemn agreement entered into so lately as last September to divide the same with another party, according to a plan laid down and marked on a map. ...

We could point our finger to the Chairman of the Board of Directors, whether the facility of attendance to parties that possessed local knowledge is not likely to prove a wholesome check to any attempt improperly to exercise the power of the Company.[6]

Norwich station, 1845

The *Norfolk Chronicle* was not the only party to accuse the railway company of subterfuge. When Yarmouth Town Council met next a report of the Parliamentary Committee proceedings was discussed. The Council were concerned on two counts in particular. Firstly, the proposal to build a bridge over the River Wensum in Norwich, which would destroy the prospect of the plan to make Norwich a port, and the proposal for a new bridge over the Bure. It was clear that the railway company's actions were arousing suspicion if not alarm:

MR. T.O. SPRINGFIELD said, he would explain the intentions of the committee ... He continued in a long and forceful speech, to support the views taken by the Committee, and contended, that the erection of the proposed bridge over the Wensum, would destroy the navigation to Norwich, whilst the projected bridge over the Bure, would prevent that river from being used, as it was now, as a harbour of refuge. The Bure was the only shelter for vessels in tempests; and he had been told, by a gentleman, that he had seen between 200 and 300 vessels stowed at its mouth. He said the Council could not allow their navigation to be destroyed, for the advantage of a private company, though they were willing to encourage them to come to Norwich, and contended that the company had not acted in good faith, as they had not, at first given any notice of their intention to apply for leave to build this bridge.

A.A.H. BECKWITH, Esq., took the same view of the matter as Mr. Springfield. ... He hoped the committee would be very watchful of the proceedings of the company. One thing in particular ought to put them on their guard; since the amalgamation of the Eastern Counties & Northern and Eastern Railways, all packages were charged at 6d. higher than they were before; that was a fact of monopoly, and if some clause were not introduced into the bill, regulating the charges, they would be always increasing.[7]

A battle for public opinion in the town between the railway company and the Cory family now began:

THE RAILWAY AND THE BRIDGE

Most of our readers will be aware, that one of the objects of the Railway bill now before the House of Commons is, to obtain leave to erect a bridge across the Bure, opposite the Lime Kiln wharf. In order to enlist the public of Yarmouth, if possible, in favour of this project, a large placard has been issued, shewing the rate tolls empowered to be levied by Mr. Cory's Act, over the suspension bridge, and promising to make the intended new bridge free to the public. — Mr. Charles Cory, one of the proprietors of the Suspension Bridge, who was absent from town at the time the anonymous placard was issued, on his return home published a temperate and manly reply, in which he shewed that the tolls allowed by the act and which were charged with being excessive, had never been levied and that the handbill had been issued to induce the inhabitants to sign a petition in favour of the bridge contemplated by the company. He then shewed that "the Railway company for valuable considerations, before the passing of their bill, entered into, and signed an agreement, with the owners of the present bridge and ferry *not to build or procure to be built, any bridge over the river Bure without the consent of such owners*; and they inserted a clause to the same effect in their act," while now they sought to wrest the property of the owners of the bridge from them without any clause as to compensation. Passing over any private reasons he might have for opposing the present bill, Mr. Cory, having shewn the evil which must ensue to the river craft by destroying the harbour of refuge so essential to the safety of craft navigating the Yare and the Bure, offers, on behalf of himself and brothers, to erect the new bridge, which, by the act of parliament, should be constructed this year (*if sanctioned by the Town and Haven Commissioners*) across the river directly opposite Fuller's Hill, a site extremely convenient to the town, as being in direct line with the Market-place, and which would leave the harbour of refuge and navigation uninjured. He then offers to meet the inhabitants of the town and the gentlemen of the Railway company at a public meeting to discuss the merits of the different pro-positions.

The result of this address has been, that on Saturday evening a meeting of the principal owners and occupiers of property in the Market Ward was held at the Angel Inn, G. D. Palmer, Esq., one of the Haven and Pier Commissioners, attended; and after the question had been well canvassed it was unanimously decided, that a bridge opposite Fuller's Hill was the best possible position that could be chosen for the interests, not merely of the Market Ward, but of the town generally.

A deputation from the meeting consisting of Wm. Johnson, (late a director of the railway company,) and Rich. Ferrier, Esqrs., was appointed to wait upon the only Yarmouth director, (Sir E. H. K. Lacon, Bart.,) to request him to communicate with his brother directors, to endeavour to obtain an object so desirable to all parties, and one which would put an end to all opposition. The hon. bart, being absent from Yarmouth the deputation waited on the secretary, (Mr. Tootal) who appeared to think the company could not have much objection to the arrangement. We understand, that the buildings, which would be required for the approach to the bridge have already been purchased, we presume, on behalf of the Railway company. Should this arrangement be entered into it will meet the views and wishes of, we believe, all parties here, would be exceedingly convenient to the Railway company as the west foot of the bridge would be directly opposite to their terminus, and there would be little difficulty in diverting the end of the Acle turnpike road so as to obviate the necessity of having two bridges.[8]

We learn however, from the Norfolk Chronicle of 16 March, that Mr Johnson's deputation "was not, we understand, very satisfactory". This was perhaps not surprising, and certainly did not come as a surprise to Charles Cory. Indeed, the "valuable considerations" he mentions was that he had agreed to purchase 100 shares in the company. It was these shares that gave him the opportunity, the previous month, to attend with his brother, Rev. William Cory (who had probably purchased his one share in order to attend), the

George Stephenson Esq., civil engineer

half-yearly meeting of shareholders. At this meeting he challenged the directors of the company about their decision to renege on their commitment:

G. STEPHENSON, Esq. was called to the Chair.

. . .

Mr. C. Cory next drew attention to certain part of the report, recommending the directors to go to parliament for the purpose of obtaining a bill to make a bridge across the river Bure; and asked the chairman, if he was a party to advising the company to make a bridge against the wish of other parties concerned.

CHAIRMAN. — Yes, I am.

Mr. C. CORY. — Then he should read part of an agreement which the company had made with Mr A.T. Cory and his brother, by which they bound themselves not to make a bridge over the river Bure, for certain considerations given, without the consent of Messrs. Cory; and the clause in the Company's Act, restraining them from so doing.—

Mr C. Cory read the following from a printed paper.

That the Company shall not erect, or procure to be erected, a bridge over the River Bure, without the consent of the owner or owners for the time being of the Suspension-bridge, and shall introduce into their bill a clause restricting them from so doing: but the said I. P. Cory and A. T. Cory agree to keep the said bridge and the approaches thereto in good and sufficient repair; and that the same shall at all times be a width adequate for the accommodation required by the traffic for passengers, goods, cattle, and carriages to and from the said railway.

Extract from Clauses inserted by the Company in their Act of Parliament, the 5th Victoria. CCXXXVII.—"Whereas A.T. Cory is, or claims to be, the owner of a certain Suspension-bridge over the Bure at Yarmouth aforesaid, and a certain ferry over the same river" "And it has been agreed between the said A. T. Cory and said Company, that the Company shall not erect, or procure to be erected, any bridge across the said River Bure aforesaid." "Be therefore enacted, that the said Company shall not erect, or procure to be erected, any bridge over the said River Bure without the consent of the said A.T. Cory, or other the owner or owners for the time being of the said Suspension bridge; and that nothing herein contained shall prejudice or affect the right or title of the said A.T. Cory, his heirs, or assigns, to the said Suspension-bridge, or to the said Ferry, or to the Tolls payable to him or them respect thereof respectively."[9]

The meeting then turned even more acrimonious when Charles Cory proposed a contentious motion to the meeting:

"That, as the erection of a bridge over the River Bure opposite the Limekiln Walk at Yarmouth, will seriously obstruct the navigation of that river and injure that of the other streams; and will in consequence incur strong opposition, of the Haven Commissioners, and the inhabitants of Yarmouth and Norwich, severally, and that as the Company, their Directors George Stephenson, Charles Tyndale, John Laurence, and Adam Duff, Esqrs., have for bona fide considerations, entered into an agreement not to erect a bridge over the River Bure, without such consent mentioned in such agreement, and which restriction is also contained in the Company's Act of Parliament, and such consent has not been obtained; it is the opinion of the shareholders, that such extension into Yarmouth should be abandoned, as such an attempt will be a useless expenditure of funds, and a direct violation of their faith."

The TREASURER said, there was only one point on which he wished to say a few words. Mr. Cory had accused the company of a breach of faith. — Now it must be quite clear to a man of business, that if when he applied to parliament to protect his bridge, he proved that power could empower another party to build one. He went to parliament and wanting to have tight hand over the company, obtained a clause restricting them from building a bridge. If he could prove to parliament that the intended bridge would impede, injure, or obstruct the navigation, they would not have a leg stand on: but if the company were enabled to show, that the bridge would not obstruct or impede the navigation, that it was necessary to complete the railway works, and that they were ready to make compensation to Mr. Cory for any injury he might legally sustain, then he must say that to accuse the company of a violation of faith was too harsh an expression. They were ready to make Mr. Cory compensation for any loss he might suffer by the erection the bridge, and thought it would have been better to have settled the amount before they went into committee; if Mr. Cory would come forward and name a sum, he had no doubt but that they would able to agree.

Mr. J.E. LACON said, an offer had already been made by the company.

The Rev. R.W. CORY said, there had been an offer made, but it was founded on the estimate of the three past years, without any reference to the prospective value.

Mr. G. P. BIDDER said, he thought it would better become the Rev. Mr. Cory, having seconded his brother's resolution, and being the representative of one share, to have been silent.[10]

By the time the bill went to Parliamentary Committee on 25 March, there were a number of bodies preparing opposition[11]. The evidence was presented to the Committee over two days during which time was given for both parties to

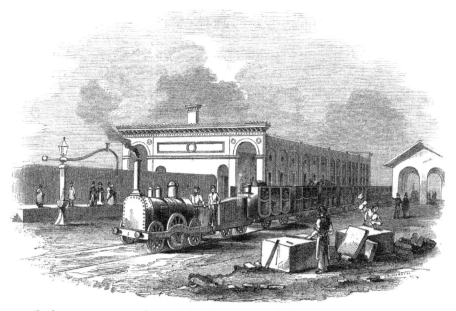

Cambridge Station, 1845, with a typical Stephenson locomotive, as also used by the Yarmouth to Norwich railway.

negotiate, but when no agreement was reached, on 27 March, they inevitably sided with maintaining Cory's rights over the crossing. Mr. Cory decided to widen the bridge by adding a footpath on either side anticipating the additional pedestrian traffic and in keeping with his side of the original agreement with the railway company.

The railway opened with great fanfare on 1 May 1844. The railway company, in defiance of Cory's rights, immediately started offering free passage to and from the Brick Quay to the railway station by steam boat! Towards the end of May 1844, Charles Cory decided to take legal proceedings against the Directors of the company[12]. The loss of revenue was small[13], and Charles Cory must have known that the railway would eventually gain the upper hand of public opinion in the town as it brought untold benefits to the townpeople. As Mr Sergeant Byles, the barrister employed by Charles Cory at the subsequent trial pointed out:

> ... and whilst, on the one hand, the advantages of the rail to Yarmouth were undoubtedly great, so on the other hand, it was certain that great improvements became exceedingly dangerous, where the right of private property was not respected. He would shew, beyond all possible doubt, that the property which the plaintiffs had bought and paid for, and in the improvement of which they had expended large sums of money, was seriously and dangerously invaded, and was in imminent risk of being destroyed, unless the plaintiffs should succeed that day in teaching the company, however powerful that body might be, that they were not with impunity to convey passengers in contravention of an act of parliament.[14]

While he could win the battle, he could not win the war. What it particularly shows is how difficult the relationship between Cory and the Directors of the railway company must have become. As the *Norfolk Chronicle* put it:

> This question now remains for the decision of a jury, and a large additional sum of money must be expended before it is decided. There can be no doubt, that though the legal right on which the existence of the bridge is founded demands to be respected, yet the bridge itself is an annoyance to the public; and we think it is a great pity, that instead of throwing away money in law, the whole question cannot be amicably settled between Mr. Cory and the Railway Company. Is it too late to effect this desirable object?[15]

Indeed, the railway company spared no expense in defending themselves, retaining both the Attorney General[16] and Mr. Fitzroy Kelly, soon to become Solicitor-General,[17] to make their case at the Nisi Prius Court on 3 August. However, the jury never got to deliberate the rights and wrongs of the case as, some hours in, the judge in exasperation asked, "Is there no mode of settling this matter?", to which Fizroy Kelly replied, "The company have always been willing to meet the case fairly."

A consultation then took place with a determination to obtain terms, if possible; and after two hours' deliberation, this very important, and in many respects difficult, case was brought to a termination, and the jury discharged from giving a verdict.[18]

The compromise reached was that the railway would get their way and be allowed to build a new bridge at their expense. However, it was Mr. Cory that would take all the tolls from the new bridge with the railway guaranteeing a minimum of £600 revenue per year from those tolls ie. if there was a shortfall, they would make up the difference. Cory agreed to reduce the walking passengers' tolls over both bridges to one halfpenny, or one penny for a day ticket. It was agreed that Mr. Cory could, "do as he pleases with the present bridge".

The Solicitor-General,
Mr Fitzroy Kelly M.P., 1845

Matters could now proceed in a gentlemanly and professional manner and both parties had to work together to the benefit of the town. Ironically Charles Cory and the directors of the railway company were meeting in London the day the bridge collapsed to discuss arrangements for the new bridge[19].

We do not know how the collapse affected Charles Cory on a personal or financial level. He did as much as he could for the sufferers by providing considerable funds[20] to meet their needs. He was as transparent as possible with the investigation held by James Walker into the causes of the collapse[21], and beforehand had offered £50 towards the expense of an impartial engineer[22] as well as making his own request to the Home Secretary[23] to appoint one.

What we do know is that soon after the disaster the Cory family's involvement with the bridge ended. On 16th May, the railway company agreed to pay Charles Cory an annuity of £800 per year, redeemable on payment of £24,000, in lieu of all charges for their passenger and goods traffic passing over the River Bure[24]. Soon after, they purchased the bridge, ferry and the surrounding land from him as well[25].

[1] Readers should not confuse this with the line running from Norwich via Acle to Yarmouth. The Yarmouth and Norwich Railway is the line running through Cantley and Reedham. It then went over the Reedham marshes, "presenting a view of Caister on the left; Burgh Castle, and various reaches of the river on the right … the line makes a curve over Breydon and ends near the Suspension-bridge at Yarmouth." A description of the line can be found in the Norfolk Chronicle — 4 May 1844, p.4. Trains still run on the line today.

[2] The Directors of the company included some powerful Yarmouth men among them Sir Edmund Henry Knowles Lacon, 3rd Baronet, a brewer and banker in the town.

[3] "A Bill to amend and enlarge some of the provisions of the Act authorising the construction of the Yarmouth and Norwich Railway and to authorise the construction of certain new Works in connexion therewith. 1844"

[4] Norfolk Chronicle — 2 March 1844, p.3. — "Our fishing merchants, and all who are connected with the fisheries, are quite alive to the importance of obtaining the most direct communication from the Jetty to the Railway terminus, without the obstruction which a toll would occasion and a petition in favour of the bill has been numerously and most respectably signed." From the account we learn that the petition, while asking for a bridge free from tolls stated, "at the same time duly estimating the rights of the proprietors of the present suspension bridge; and that the railway company may be made to pay a fair and equitable compensation, to which they deem the parties quite entitled."

[5] Norfolk Chronicle — 9 March 1844, p.3.

[6] Ibid., 4.

[7] Ibid., 5.

[8] Ibid.

[9] Norfolk Chronicle — 17 February 1844, p.3.

[10] Ibid.

[11] Norfolk Chronicle — 23 March 1844, p.3. — "Several gentlemen of the long robe have received retainers to conduct opposition from the Town Council of Norwich, the Dilham Canal Company, the Haven and Pier Commissioners of Great Yarmouth and the Proprietors of the Suspension Bridge over the river."

[12] Norfolk Chronicle — 25 May 1844 p.3. — "A rumour is very currently in this town, that Mr. Cory has taken legal proceedings against the Directors of the Railway for injuring his right of toll from Yarmouth to the Fen Farm over the river Bure by carrying passengers from the Quay to the Terminus. No proceedings had been commenced on Saturday; Mr Cory is, we believe, in town for the purpose of obtaining opinions as to the best course to be pursued."

[13] Norfolk Chronicle — 8 June 1844, p.3. — "Cory v Yarmouth and Norwich Railway Company. — In the Vice-Chancellor's Court, on Saturday, June 1st, his honour [Sir J. Wigram], refused to grant the interim injunction in this case, to restrain the defendants from carrying their passengers by boat across the Bure, observing that the injury which could result to the company even if the injunction were granted, must be very small; but as the company would undertake to file their affidavits by Wednesday next, when the motion could be fully heard, and the highest estimate the plaintiffs gave of their probable loss was £3 a day, it

would be sufficient, in the mean time, to require the defendants to keep account of the passengers who should cross the river in their boat, and abide such order as the court might make for compensating the proprietors of the suspension bridge."

14 *Norfolk Chronicle* — 10 August 1844, p.6.
15 *Norfolk Chronicle* — 15 June 1844, p.3.
16 *Norfolk Chronicle* — 22 June 1844, p.2.
17 In March 1845 Mr Fitzroy Kelly defended John Tawell, "the Quaker murderer" in one of the great poisoning crimes of the early Victorian period. This case became famous for being the first in which a criminal was apprehended using the electric telegraph. John Tawell was born in Aldeby in Norfolk and worked as a shop assistant in Great Yarmouth for over ten years before moving to London. Convicted of forgery he was sentenced to 14 years transportation to New South Wales where he eventually made his fortune. On returning to England he began an affair with his housekeeper. On the 1st January 1845 he poisoned her with prussic acid (cyanide). Tawell was publically executed on 28 March 1845. Fitzroy Kelly's argument that Sarah Hart had eaten too many pips (pits) of her apples and got poisoned by the prussic acid in the pips led to the nickname "Apple-pip" which followed Kelly for the rest of his life. See - http://www.johntawell.com/.
18 *Ibid.*, 14.
19 *Norfolk Chronicle* — 10 May 1845, p.4.
20 *Ibid.*, Ch 2, 5.
21 *Bury and Norwich Post* — 21 May 1845, p.3. — "The Coroner proceeded to say that Mr. Cory had, in the handsomest manner possible, said that every document relating to the bridge, from its origin to the present, should be at the service of the jury. There were the original specifications and the agreement with the party building the bridge, there were surveys of a later date, the agreement between the Railway Company and himself, to which Mr. Stephenson was privy. Mr. Cory was anxious to have the matter fully gone into and would render to the Engineer and Jury every assistance in his power to their investigation of the cause of this melancholy accident. He thought it right to make these observations in justice to Mr. Cory.

> "Several Jurymen remarked that Mr. Cory's conduct, from the commencement, the sympathy he had manifested towards the sufferers, and the readiness with which he proffered his aid to further inquiry, was very handsome, and did him much credit."

22 *Royal Cornwall Gazette* — 16 May 1845, p.1. — "In the course of proceedings, it was stated by a juror, that he was authorised to say on behalf of Mr. Cory that he was anxious to have the late bridge examined by any competent engineer, to be appointed by the Coroner, towards the expense of which he offered 50l."
23 *Norfolk Chronicle* — 17 May 1845, p.3. — "We have this morning ascertained that Mr. C. Cory has applied to Sir James Graham, requesting him to send down a government engineer …"
24 *Norfolk News* — 30 August 1845, p.1,3. — Report of the Half-yearly General Meeting of the Proprietors of the Norfolk Railway Company —"It appeared to the directors most desirable to put an end to these disputes, which were in a constant source of litigation and ill-will;".

  25 *Norfolk News* — 14 June 1845, p.3. —

> "The public will probably be glad to learn that an agreement has at length been

entered into between Mr. Cory of Yarmouth, the proprietor of the Suspension bridge, and the owner of the right of ferry over the Bure, at the North Quay, and the Norwich and Yarmouth Railway Company, the latter having purchased the property, (i.e. the bridge and ferry, with the Vauxhall gardens, and some adjoining meadows, including Padgett's brew office,) for the sum of £26,000. This will give the Railway Company the desired facilities at the Yarmouth Terminus, and no doubt a free passage over the river for railway passengers will follow."

The history of the rail bridge that was subsequently build over the Bure at Lime Kiln wharf can be found here: http://www.greatyarmouthpreservationtrust.org/media/uploads/History_of_Vauxhall_Bridge.pdf (last accessed 25 March 2015).

7

Circus and Clown

"Dear Public, you and I of late,
Have dealt so much in fun;
I'll give you now a monstrous great
Quadruplicated pun —

"Like a grate full of coals I'll burn
A great full house to see;
And if I am not grateful too;
A great fool I must be."

Poem on a poster advertising Nelson's stunt on the River Bure, 2 May 1845

In 1768 Philip Astley, a former Sergeant-Major in the 15th Light Hussars, established a riding school on a fenced wayside field on Lambeth Marsh. Styling himself as the "English Hussar" he performed feats such as straddling two cantering and jumping horses, doing headstands on a pint pot on the saddle and a "parody of riding by a foppish tailor". He would charge 6d admission or 1 shilling for a seat. Astley never referred to his entertainment as a circus, but called the arena, a "circle" or "amphitheatre". His business soon grew and by 1777 he had a large wooden building by Westminster Bridge. Astley's Amphitheatre defined the form of the modern circus[1] and others soon realised the potential of this new and lucrative form of entertainment.

A show consisted of daring equestrian acts and "pantomimes" based on popular tales and events of the time padded out with clowns, jugglers or acrobats. Astley, and those that followed him, would show their daring equestrian skills to the full by exploiting suitable stories. One particularly popular adaptation was "Mazeppa; or the Wild Horse of Tartary". Originally a folk tale, and made famous in the period by a poem by Lord Byron, it was first adapted for the stage in 1831 by Henry M. Milner[2]. It tells the story of a Ukrainian warrior who travels to the Polish court and falls deeply in love with a Countess, who is married to a very much older man. He is discovered and punished by being sent back to the steppes

Astley's poster featuring Mazeppa.
British Library.

of Ukraine strapped naked to the back of his white charger.

The Cooke circuses were close to unique in that William's father, Thomas Taplin Cooke, had produced nineteen children and therefore, when it came to performers, they employed few outside the family. While the occasion of the disaster was William's first visit to Yarmouth, Thomas Taplin or maybe his grandfather, Thomas[3], had performed in the town before[4].

Thomas, the founder of this circus dynasty, was an accomplished rider, acrobat and ropewalker. In November 1806 he fitted up a former iron foundry on the Southside of the Seagate in Dundee and called it "Cooke's Olympic Circus". The next year he opened his "New Olympic Circus" in Virginia Street, Aberdeen. The business and family grew, and by Saturday 18 September 1830 his son, Thomas Taplin Cooke, and the whole family gave a special Royal Command Performance at the Royal Pavilion, Brighton. From that moment on, he and his descendants were able to include the word "Royal" in their circuses' titles.

Thomas Taplin's circus was also based in Scotland and like Astley's Amphitheatre in London was a wooden, temporary structure in Glasgow. Today one imagines a big-top tent as being the basis of a touring troupe, but 19th century circus proprietors either took a pre-fabricated wooden building or would build a new one in-situ if performing in a town for several weeks. In September 1836, Thomas Taplin announced that he was to take his circus to America and chartered the *Roger Stewart* to transport his 35 horses and 70 performers with their wives and children to New York. The circus performed in New York, Boston and Philadelphia. In December 1837, the company moved on to Baltimore. It was here that a major setback befell them when, on 8 February 1838, the building was burnt to the ground and, although no-one lost their life, fifty-two of his horses were killed. With no insurance, he had lost everything and was ruined. He attempted to carry on with untrained American horses, but eventually had to return to Britain that year.

William Cooke, his second son, had gone with the retinue to America and together with his brothers began to re-build their father's business. In 1839, a "large and elegant" circus was built on a vacant piece of ground behind the York Hotel in Edinburgh[5] and performances began. This was followed by seasons in Dundee, Glasgow and Edinburgh again. In late 1840 they built a circus in Hull but returned to Edinburgh in 1841. It was during this Edinburgh season that the first mention of Nelson as a performer in Cooke's circus is made. On Thursday 24 March 1842, the first 200 people to the pit and gallery were to receive a full-length portrait of him.

In April the company returned to Aberdeen and we get the first mention of Nelson's stunt with geese:

> In 1842 Aberdonians were again glad to welcome back Mr Thomas Taplin Cooke. He opened Cooke's Royal Circus in Union Street during the month of April. ... Mr William Cooke who was manager of the circus this season gave riding lessons to those that wished them ...
>
> On 16th June Mr Nelson, one of the clowns of the circus, created quite a sensation by a novelty he introduced on the date of his benefit. He announced by handbills that he was to navigate the harbour from the lower basin to Regent Bridge in a tub drawn by two geese. This had the effect of exciting the curiosity of an immense crowd; many thousands of both sexes lined the quays long before the hour of the exhibition at half past 4 p.m. At an early hour respectable tradesmen, shopkeepers, learned lawyers, pious divines, sober matrons and prude old maids, and a large number of boys and girls, might have been seen making their way down the quays from all quarters of the town, and from every avenue to see this wonderful sight. All bridges and spots where a good view could be obtained were one black mass; no launch of the largest of our ships from Aberdeen harbour ever

saw such a crowd as that day turned out to see the clown and his geese. When Mr Nelson arrived dressed in the costume of a merry-Andrew, and driving the geese before him with a long stick, he informed the spectators he would be unable to go the route he proposed, as the wind was contrary. However he sailed up the river in a boat, and let out from the bridge. He thus satisfied the curiosity of his mass multitude, and the advertisement had the effect of filling Cooke's Circus in the evening.[6]

A performer's "benefit" meant that he or she would get the profit that night and was standard practice to boost an individual's earnings. Although others had done the stunt[7], Nelson's experience in Aberdeen showed him that for the cost of a few handbills and posters, and an hour's free entertainment, he could fill his benefit evening to capacity. In the next few years, the geese "trick" became Nelson's signature when it came to drumming up a crowd for his benefit night. Indeed, one might suspect that it was almost expected of him in the places where the geography was suitable. However, it was not without problems. On 22 May 1844, Nelson performed the stunt from Park wharf to Broadwick's wharf, on the canal at Kidderminster where, "the crowd who were assembled threw pieces of bread to the geese, to divert their attention, and some threw stones and sticks at them. This, of course, rendered the task more difficult, but it was performed by Mr. Nelson in beautiful style.[8]"

Interference could be an occupational hazard and, if the crowd believed they had been misled they made their feelings known. Barry the clown (who was also renowned for the same stunt) and was on contract with Nelson, and a third clown called Twist, at Astley's Amphitheatre found this to his cost in October 1844:

> A STALE JOKE. — Old father Thames was on Friday again the scene of much bustle and excitement, Mr. Barry, one of the clowns at Astley's Theatre, having a second time announced he would sail from Vauxhall to Westminster-bridge in a washing tub drawn by four geese. To add to the attraction of this strange feat it was stated that Mr. Carter, the "Lion King," would follow the clown in an open boat with one of his beautiful tigresses. In the latter respect, however, the spectators were disappointed, for no tigress was to be seen. By an alteration in the plan originally agreed upon thousands of persons were deceived. On account, it was said, of the tide, the clown sailed from the Red-house, Battersea, to Vauxhall-bridge, instead of from the latter to Westminster-bridge. The crowd was on this occasion much greater than on the previous occasion. The whole distance from Westminster-bridge to a spot opposite the Red-house was lined with people, while an immense number collected on the wharfs on the

Barry the Clown on the Thames, 1844

opposite shore. At 5 o'clock Mr. Barry embarked in his "frail vessel," the geese being regularly yoked to it in shafts, and thus he proceeded driving "four in hand," amidst the laughter and cheers of innumerable water-parties preceding or following him, and the music of a large brass band playing "Rule Britannia", and "See the conquering Hero comes". Some mischievous fellows caused the buffoon some uneasiness by endeavouring to "foul" him and his geese, that is, to run against him with their boats, but the clownship was on the look out for such attacks, and escaped without injury. At Vauxhall he got into a common boat and rowed to Westminster, where the crowd, expecting to see him arrive with his tub and geese gave vent to their displeasure in that kind of hissing vulgarity termed "goose," and Mr. Barry got into the theatre with all possible speed.[9]

The London crowds could not get enough of these exhibitions and within three days Nelson, and his companion clown Twist, for their "joint benefit", decided to conduct a race:

> The CLOWNS' SAILING MATCH. — The third of these, what at one period was considered to be an undertaking of very great novelty, that of a person being pulled or drawn down the river by means of four geese being attached to a washing tub, came off on the river, between the Red-house, Battersea, and Vauxhall-bridge, on Monday afternoon. As this was the third within a fortnight, undertaken by persons connected with Astley's Theatre, something of the novelty had declined, when for the purpose of

exciting the taste in the public, it was announced, that upon this occasion A. Nelson and Twist, the pantaloon and clown, would each make a voyage in different washing-tubs drawn by their respective teams; this announcement, accordingly drew together a large assemblage of persons at the Red-house, when, after a lapse of half an hour, the two aquaties, in full dress, namely, their theatrical costume, took their seats, and having picked up their reins, they each fired a pistol, when the medley procession started. They were accompanied by about 20 boats, which formed a kind of circle round the geese and tubs, and having gained the middle of the stream they floated down to Vauxhall-bridge which was reached by half-past five o'clock. After getting about 200 yards through the bridge, the two washing tubs were brought up to a barge, when Nelson and Twist got out of their craft, and each putting on a great coat, were conveyed to Lyon's yard, Stangate, where they were landed. On the passage down the water was very rough, and as the tubs passed by the above-bridge steamers, the surf from them caused the tubs to perform certain motions which appeared not to be quite agreeable to the heroes of the ring of the Amphitheatre. The original intention was to land at Westminster-bridge, but which was very wisely prohibited by the police as on the first of these novel undertakings, the mob in the Palace-walk was so great, coupled with the disturbances and robberies, that it was deemed advisable to prevent repetition. Whatever effect it produces for Nelson and Twist (that night being their joint benefit) it was attended with beneficial results for the proprietors of Vauxhall-bridge, as at the time the party passed beneath it, there could not have been less than 2,000 persons on the bridge, straining their eyes to catch a glimpse of the washing-tubs, geese, and buffoons.[10]

In January 1845, when William Cooke applied to the Mayor of Yarmouth "for leave to erect an amphitheatre, on the Theatre Plain[11]" he had no idea that his visit to Yarmouth would end in tragedy. He had set up a touring circus with his own family in late 1843. During that first year he had taken the troupe to Norwich, and had returned at the beginning of 1845 to an appreciative audience:

> The lovers of equestrian sports may enjoy a treat, by visiting the royal circus of Mr. Cooke at the Ranelagh gardens. We are informed the company is not to be surpassed by any in England. The juvenile performances of Miss Cooke a child three years and a half old, are very amusing. We understand the scene of St. George and the Dragon is very good.[12]

By May 1845, Nelson had performed the stunt dozens of times and there was no reason why he should not do so in Yarmouth with his benefit night scheduled for Friday 2 May. We can get a flavour of the excitement of the circus and the effect the advertisements had on a community by this autobiographical account of that 1842 Aberdeen outing:

> ... I was accompanied by several school companions about my own age. Alfred Cooke was the hero of our talk. So was Charlton the clown. Attempts were made by us to imitate his posturing. He could stand on his head, walk upon his hands, leap like a frog with body bent and legs over his shoulders; and wounds and bruises not a few were the result of our unsuccessful efforts to succeed as young acrobats.
>
> The bills which were issued of the performances were sometimes illustrated by scenes of the circle, rough wood engravings printed in red or blue and very conspicuous in colouring. At that time many merchants displayed them in their shop windows. Window dressing then was not the fine art it is now, and in many shops the windows were anything but neat and attractive. Several of the shop assistants were permitted to take charge and hang up boards with the bills pasted on them near the shop door or on the nearest lamppost, for which service they received an occasional pass to witness the performance. On the bills after the circus arrived a winged horse was exhibited, and on this, my first visit I was much disappointed that this animal, which was a novelty in Natural History, and, to me, one of the greatest inducements to be present, was not visible. The wood engraving which I have before me as I write, and which so attracted my fancy and wonder, is the figure of a prancing horse rearing on its hind legs, with the forelegs greatly elevated and pawing the air. The head was held high, the eyes large and distended, the mouth open, the mane large and flowing, the tail long, bushy, and waving. The wings are partly outspread as if the animals were to fly, and a broad bellied band encircles the fiery steed. In the distance there is a pyramid in rough outline, giving the picture an Eastern look, and portraying, as afterwards I found out, a horse with an existence as real as the lamp of Aladdin or the fairy palaces so minutely described in the Arabian Nights Entertainments.[13]

The author goes on to give his opinion of Nelson and to describe his stunt in the harbour:

> Besides Merriman Charlton, there were two other clowns, Nelson and Wells, but neither of these were of much account. Nelson was no acrobat, nor a great jester. He played sweetly on the dulcimer and other musical instruments. On 17th June 1842, he was to take his benefit. On that day by handbills and advertisements it was announced that he would

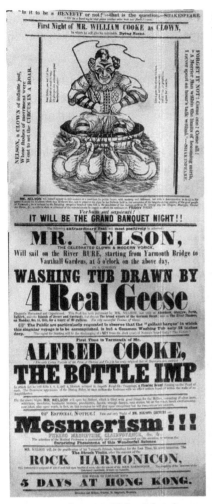

A poster advertising Nelson's benefit at Cooke's Royal Circus, May 2nd 1845

sail in the harbour in a boat drawn by two geese, and the hour of starting was announced to be 4 o'clock in the afternoon. In anticipation of this great event, thousands crowded to the quays on both sides. At the hour, Nelson in his clown dress entered what appeared to be half a barrel, to which two live geese were attached by cords. In one hand he held a cord which was fastened to the geese. It was rather slow and tedious work, and the geese were impelled to move by a long wand which he frequently applied to them. Ultimately, they reached the top of the harbour, but not before one of the geese had been done to death. To me and many others this was a wretched and disappointing performance, and the multitudes that witnessed it were, I am sure, very much more impressed by the pictures on the bills illustrating the attempt than by witnessing its performance.[14]

In Yarmouth, Nelson followed the usual routine issuing both posters and handbills, which drew on his previous exhibitions and made reference to the other "Nelson" and local hero[15].

This Feat had been performed by Mr. Nelson, not only at **Aberdeen, Glasgow, Perth, Falkirk** and the **Islands of Jersey and Guernsey** last, also on **The broad waters of the German Ocean** and on the **River Thames on Monday October 11, 1844 for a wager of 60 guineas**. *The only successful trainer of Geese.*

The Public are particularly requested to observe that the "gallant barque" in which this singular voyage is to the accomplished is but a Common Washing Tub, only 10 inches deep. The signal for Starting will be the discharging of a GUN from the deck (rim) of Nelson's Vessel (tub) "The Victory".

We learn that the first performance of the circus was on 26 March and had sell out audiences, indeed on Monday 31 March an accident had occurred during the performance, the audience was so large:

Mr. Cooke, who is now on his first visit to Yarmouth has been fortunate enough to secure a site for the erection of a commodious and convenient building on Theatre Plain. The performances commenced most auspiciously on Wednesday se'nnight, and we doubt not that the visit he has paid us, while it affords abundant gratification to those who attend, will prove encouraging to the spirited and worthy proprietor. Mr. Cooke has provided not only a spacious amphitheatre, but a phalanx of artists, (about fifty in number) with a stud of thirty horses. The performances are most attractive. Those of Messrs. Cooke and Barlow, in various characters they sustain, are unrivalled; while the chastity of the comicalities introduced are inoffensive to the most fastidious taste. The house has been filled to overflowing every evening; and we should imagine that the attendance each night has not been less than 12 to 14 hundred.

Mr. C's stay will be short, and we recommend to all lovers of equestrian, gymnastic, and pantomimic performances, to pay him an early visit. We observe tomorrow evening [Friday] Mr. Cooke intends giving the proceeds of the entertainment to the Yarmouth Hospital, when we hope to see a good attendance.

Accident — On Monday evening last, an accident occurred at Cooke's Circus, in the 2nd act of "St George and the Dragon," in consequence of the gallery being so crowded that a number of persons ventured themselves on the boarding thrown over the avenue leading to the stables; this not being sufficiently strong to bear so great a weight, broke down with a tremendous crash. We are happy to say, that only one person was seriously injured, a man named William Lilly, aged 53, who had his leg broken in two places. Mr. Cooke immediately sent for a fly, which conveyed him to the Hospital, and we are happy to state that he is, under the hands of Mr. F. Palmer, doing well.[16]

From other descriptions we know that Cooke's wooden circus buildings were lit by a large gas chandelier and given that Bridge's circus, who visited the town in 1837 and 1840, used gas to light their building[17], we can be sure that the Cooke's amphitheatre on Theatre Plain would have also done so.

By the last week of April, audiences at the circus had not dwindled and the arrival of William's brother, Alfred, made them as strong as ever.

Cooke's Circus — This most agreeable and interesting scene of

Cooke's Royal Circus at the New Standard Theatre, Shoreditch, 1845

amusement has presented a fund of attraction during the past week, which may be fairly pronounced unrivalled, and we are glad to find Mr. Cooke has not catered unsuccessfully. The circus, which is by far the most capacious ever erected in this town, has been crammed almost to suffocation; and the performances themselves have been highly creditable to the performers themselves and to Mr. Cooke. The exquisite skill and taste of Mr. G. Cooke and Master W. Barlow, in a new comic act (on two coursers), entitled "The Toad in a Hole, or a Christmas Pie," received great applause; as also did the performance of "Dog Nelson," and Signor Germani, the Italian Juggler. — On Wednesday evening, Mr Alfred Cooke made his first appearance in this town (having just arrived from Manchester with his stud of horses, previous to the company going to London). Being the first sight of "Turpin's Ride to York, or the Death of Black Bess," the arena was crowded to excess, and Mr. A. Cooke was warmly cheered at the close of the performance.[18]

Since Cooke's circus had a booking at the New Royal Standard Theatre[19] in Shoreditch for Whitsun, Nelson's benefit was towards the end of their Yarmouth run.

On the morning after the collapse of the bridge, William Cooke appeared in front of the magistrates of the town:

TOWN HALL — Saturday Morning Ten o'clock.

The Magistrates met at this hour; there was a very full bench The Mayor, W. H. Palmer, Esq, in the chair.

Mr. W. Cooke, the proprietor of the Circus, (who appeared to be labouring under great mental excitement,) proceeded to address the bench. He informed the Magistrates that he had come to the conclusion of withdrawing the company from the town.[20]

That evening he distributed the following handbill in the town:

It is with the deepest feelings of regret that Mr. W. Cooke has to announce to the gentry and inhabitants of Yarmouth, his intention of immediately closing his Circus, in consequence of the melancholy circumstance which has so recently occurred. No person can feel more deeply than he does the sad catastrophe which has filled with grief the inhabitants of Yarmouth. He trusts a discerning public will trace the calamity to its proper source — the Bridge, not being of sufficient strength to support the weight of so many persons; as the circumstance might have occurred in witnessing any public exhibition.[21]

Despite his swift departure William Cooke was to bring his circus back to Yarmouth some eighteen months later[22].

William Cooke was not the only one distressed by the catastrophe. *The Times* reporter, sent to the town, tells us on Wednesday 7 May:

It is generally stated that Nelson, the Clown, the unhappy yet innocent cause of this sad catastrophe, has been very deeply affected by the consequences of the foolish exhibition in which he took so prominent a part, and that he has suffered great mental and bodily anguish. He left Yarmouth last night.[23]

In January 1846, Nelson left for New York in order to try his luck in America. There he was to charm audiences with his Rock Harmonicon and Pine Sticks[24]. He had returned to Britain by September, and ironically his first engagement was with Cooke's Circus at the Victoria Gardens in Norwich where:

Not the least astonishing performance of the evening was the exhibition of Mr. Nelson's stick harmonicon (if we may so call it) which consisted of sundry pieces of wood, varying from six inches to a foot in length, joined together with twine or straw, on which several popular airs were played with the utmost dexterity and precision.[25]

Some considered that he should never undertake the exhibition again[26], and evidence shows that it was some two and a half years before he undertook it a further time[27]. On a poster for his benefit at the Pavilion Theatre in 2 April 1851, it states that "And now for the **205**[th] Time **Mr. NELSON will Sail in a Common Wash Tub DRAWN BY 4 REAL GEESE!** Starting from the Blackwall Pier at 3 o'clock precisely".

Despite this, the number of recorded occasions that Nelson undertook the stunt diminished over the years, but this did not stop the popularity of the spectacle wherever and whoever performed it. Some two months before Nelson reprised his stunt, another accident occurred in Sheffield where Teasdale, the clown at the visiting circus, decided to perform the "trick" using four ducks on the River Dun:

Pavilion Theatre, Wednesday 2nd, April 1851 for the Benefit of Arthur Nelson.

The Love for Novelties. — We last week noticed the announcement by a comedian, named Teasdale, of a performance at the circus, under the title of "The Chesterfield Murder." The authorities having very properly prevented the carrying out of Mr. Teasdale's intention, it appears he was not to be disappointed in his wish to treat the public to some novelty. Accordingly he announced that, on the evening of Monday last, he would be drawn in a tub, by four ducks, from the Iron bridge to the Lady's bridge. Mr. Teasdale was not disappointed in what he anticipated would be the effect of this announcement. At the appointed time, thousands of persons, male and female, thronged the bridges, banks, and avenues leading to the river to witness the performance of this feat. Mr. Teasdale, who is a clown at the circus, started with his tub and ducks from the Iron bridge. Both were, however, perfectly unmanageable. The ducks would not go as directed by him, and the tub, after rocking for some time from side to side, overwhelmed, and turned its occupant, amidst the laughter of the lookers on, into the water. All were anxious on both margins of the river Dun to witness the various mishaps which Teasdale met with; and it was truly laughable to see him pushing his boat and ducks forward, in shallow water, towards the Lady's bridge, his intended landing place. At this juncture, the scene was changed from merriment to disaster, as about twenty yards of a wall fell in a yard in the Wicker, laden with spectators, including four or five females, and nearly thirty were precipitated into the water beneath, a depth of at least ten feet from the yard. Fortunately, it was not deep enough to drown any of them, but several met with severe bruise and contusions, and were not released without considerable difficulty. In a yard, at the tilt, near Lady's bridge, a young man, a journeyman tailor, met with a shocking accident. There were some wooden palisadings in this place, surmounted with spikes, and he, to save his life from the rush which took place, placed both his hands upon them, and, at this moment, the rush to peep at the foolish feat was still greater, and his left hand was perforated by one of the spikes, which passed between the fore finger and went out near the wrist. One of the fingers of the other hand shared a similar fate, and it will be a miracle if he recovers the use of his right hand again. So much for "Tomfoolery."[28]

In April and May 1858, the clown at Ginnett's Circus, Dan Cook, performed the event three times, at Lincoln, Wisbech, where "it was estimated that not less than six thousand persons were assembled (more than half the population of Wisbech)[29]" to watch, and in Spalding where the "police were recommended not to allow the Victoria Bridge to be over crowded on Tuesday next, when Ginnett's clown would pass down the river in a tub drawn by geese, in order to prevent any accident occurring at that Bridge.[30]"

By the late 1870s the stunt had lost its appeal and, while it was performed at the odd regatta and other events where crowds were already present, it was never again to have the impact it once had.[31]

[1] See Rendall, M. (2014) *Astley's Circus: The Story of an English Hussar*.

[2] Henry M. Milner was a playwright and author of melodrama and popular tragedies. His most notable work was "*The Man and the Monster; or the Fate of Frankenstein*" which opened in July 1826, six months after Mary Shelley's book, "*The Last Man*" was published. He adapted Byron's 1819 poem in 1831.

[3] McMillian S. (2012) *Cooke's: Britain's Greatest Circus Dynasty* – Thomas and Thomas Taplin Cooke are often confused as many advertisements, posters and newspaper reports of the period refer to Cooke's Circus, Thomas Cooke's Circus or Mr Cooke's Circus; because both had the same forename it could mean either father or son.

[4] *Norfolk Chronicle* — 1 Apr 1820, p.3. — "Ladies and Gentlemen of YARMOUTH and vicinity are most respectfully informed, that Mr. COOKE has fitted up, on the THEATRE PLAIN, an elegant and commodious CIRCUS, which he intends Opening on Friday, April 7th, 1820.".

[5] *The Scotsman* — 2 November 1839 — cited in McMillian S. (2012) Cooke's: Britain's

Greatest Circus Dynasty.

6   *The Northern Figaro* — 4 June 1898 in an article, "Circuses in the City from 1807 to 1897 by Harry S. Lumsden" cited in McMillian S. (2012) Cooke's: Britain's Greatest Circus Dynasty.

7   *Leeds Mercury* — 18 July 1818, p.4. — "Singular Wager. — On Thursday week, Mr. Usher, the clown at Cobourg Theatre, undertook to go from Blackfriars-bridge to Westminster-bridge in a washing tub drawn by four geese. He started the tide at half-past one, and accomplished this singular wager in one hour. An immense number of persons witnessed the undertaking: after completing it, he sailed to Cumberland-gardens, and there offered, for a wager of 100 guineas, to return thence through the centre of London-bridge; but no person would accept the challenge.".

8   *Bristol Mercury* — 25 May 1844, p.6.

9   *West Kent Guardian* — 19 October 1844, p.2.

10   Ibid. — With regard to the last remark, the Vauxhall bridge in London was a toll bridge and would have charged the people watching to cross. In the case of the Yarmouth Suspension Bridge, the toll-house had been moved in 1837 (see chapter 5) meaning that access to the bridge could be had from the Yarmouth side without paying the toll.

11   *Norfolk Chronicle* — 1 February 1845, p.3. — "Cooke's Circus — Mr Cooke applied to the Mayor on Monday last, for leave to erect an amphitheatre, on the Theatre Plain, Mr. Cooke stated, that he purposed coming in about a month, and that he wished to stay five weeks. The Major granted him leave, but said he must limit the period of exhibition to a month.".

12   *Norfolk News* — 18 January 1845, p.3.

13   *Bon-Accord* — 21 May 1903, in an article, "Cooke's Circus: My First Visit" — A.S.C. A.S. Cook wrote a book entitled, "*Aberdeen Amusements Seventy Years Ago*", in 1911, where he recounts the same story — cited in, McMillian S. (2012) Cooke's: Britain's Greatest Circus Dynasty.

14   Ibid.

15   Admiral Lord Nelson landed in Yarmouth on 6 November 1800 to a hero's welcome and was given the Freedom of the Borough. At his speech from the balcony of the Wrestler's Arms he said, "I am a Norfolk man and I glory in being so!". He returned on 1 July 1801 following his victory over the Danes at the Battle of Copenhagen.

16   *Norfolk Chronicle* — 05 April 1845, p.3.

17   *Norfolk Chronicle* — 25 July 1840, p.3. — "The place is brilliantly illuminated with 70 gas lights, supplied by the Yarmouth Gas Company, which has very recently laid down a new main one-third larger than the old, by which means they are able to furnish a more adequate supply of light than before.".

18   *Norfolk Chronicle* — 26 April 1845, p.3.

19   *The Era* — 18 May 1845, p.2. — "STANDARD. — The Whitsun visitors to this very pretty little theatre were entertained with an entire change of performances, the spirited managers, Messrs. Johnson and Nelson Lee, having engaged the troupe of Mr. Cooke, the equestrian, and who made their debut on Monday evening to a densely crowded house. The performances commenced with a military spectacle, entitled "The Conquest of Tartary; or, The Eagle Rider of Circassia, and her Monarch Steed of the Desert," followed by some very excellent scenes in the circle; the evenings amusements terminating with a grand entrée of twelve horses, entitled "The Warrior's Dream". The spectacle was produced

under the direction of Mr. W.D. Broadfoot, and reflects the highest credit upon that gentleman, whose judgment in such matters has been frequently put to the test, when director of Astley's Theatre. Mr. Arthur Nelson, as Clown, and T. Swan, as Buffoon, kept the company in high spirits throughout the performances in the ring. We doubt not that the speculation of Messrs. Johnson and Lee will turn out considerably to their advantage." – Johnson and Nelson Lee had only just taken over the licence of The Royal Standard and began organising circus and equestrian shows to take place during the summer season (May 15 to September 15). These additional shows provided revenue for the regular season, which ran the rest of the year. Nelson Lee was also to become a prolific pantomime writer, which kept clowns like Nelson, who moved between theatre and circus, in constant work. As Lloyd's Weekly put it in December 1849, while summing up the pantomimes in the city and provinces, "… Nelson Lee, who has a finger in every pie of every theatre of note …"

[20] *Norfolk News* — 10 May 1845, p.2.

[21] Ibid., p.3.

[22] *Norfolk News* — 5 September 1846, p.3. – "COOKE'S CIRCUS we understand will be opened on Thursday and Friday evenings, with an excellent company of equestrians, a noble stud of horses, and beautiful fairy ponies." Although he is not mentioned, it is likely that Nelson was in Cooke's retinue since he certainly performed in Norwich the following week, see note 25.

[23] *The Times* — 7 May 1845.

[24] *The Herald* — 3 April 1846 – "The soul-stirring melodies induced by Arthur Nelson ... are truly overpowering sweet and enchanting, and while they delight the ear, [they] surprise the mind, which in consideration that music like this can be wrought from simple, rough stones, cannot but conclude that this age is in reality the age of wonders." cited in Strong on Music: The New York Music Scene in the Days of George Templeton Strong Volume 1 Resonances 1836-1849 (1998) Lawrence, Vera Brodsky p 414-415.

[25] *Norfolk News* — 12 September 1846, p.3.

[26] *Bury and Norwich Post* — 7 May 1845 p.3. – "Surely, surely, it is the last time such a ridiculous, so degrading a feat, will be attempted."

[27] *Yorkshire Gazette* — 25 September 1847 p.6. – It took place at Stockton-upon-Tees from Stockton Stone bridge down the river Tees. The event was so popular it was repeated five days later – "The banks on both sides were crowded to see this sight, and the question remains to be solved whether the people on land or the birds in the water were the greatest geese".

[28] *Sheffield and Rotherham Independent* — 31 July 1847 p.8.

[29] *Lincolnshire Chronicle* — 30 April 1858, p.3.

[30] Ibid.

[31] *Edinburgh Evening News* — 11 June 1877 p.4. –
"THE ACTOR, THE GEESE, AND THE TUB.
> Saturday afternoon Mr. Felix Rogers, of Sanger's Amphitheatre sailed in a tub accompanied by four geese from Battersea to Westminster Bridge. The tub was two feet deep by two feet six inches in diameter, and was balanced by heavy weights, four geese being harnessed at the front. The actor was dressed in a naval captain's attire, and sat on a seat fixed across the tub. The start took place at twenty minutes

to two on the ebb tide, which was flowing strong enough to carry the tub steadily along, the geese appearing to do little or nothing towards drawing it, their heads being often turned towards the actor. It was a few minutes after three when the tub passed under Westminster Bridge, and it was carried as far as a large timber wharf on the Surrey side before a landing could be effected. After some delay Mr. Rogers was got safely into a boat and rowed to the Westminster Bridge steps, where he landed amid the admiration of about 30 small children and a sprinkling of spectators on the bridge.".

8

Legacy and Memory

Deep Waters could not douse
the enduring flame
that burned for those thus blighted
Now let them stand as one,
properly remembered
and in part righted

Poem written by Ian Nimmo White on the memorial erected to the victims of the Tay Bridge Disaster of 1879, unveiled on 28 December 2013.

Neither of the most horrific Victorian bridge disasters had a memorial to their victims until 2013. Of the two, The Tay Railway Bridge disaster is the most remembered. On 28 December 1879, 59 people lost their lives when the bridge from Dundee to Fife collapsed, sending a passenger train plunging into the Firth of Tay.

A contemporary image of the Tay Bridge disaster.

The Tay and Great Yarmouth memorials were the result of the unstinting efforts of two campaigners with similar motives. Both felt that little had been done to honour the dead at the time and that it was now time to do so.

Ian Nimmo White, a Leslie poet, established the Tay Rail Bridge Memorial Trust in 2010 in order to raise money to erect a memorial to the victims of the disaster. In fact twin memorials were erected either side of the river. At the ceremony on 28 December 2013 he stated his reasons:

> This is a 134-year oversight which is finally being righted. The families weren't treated all that well by the railway companies after they lost their loved ones. And the rail bridge was an injustice in itself because it was an accident waiting to happen.
>
> But this is an opportunity to give the families some kind of payback.[1]

Three months earlier the memorial to the victims of the Yarmouth Suspension Bridge disaster had also been unveiled. Julie Staff, the owner of a deck-chair business in Yarmouth and campaign fund-raiser for the Yarmouth memorial said:

> Anybody who came will have played a part in the telling of the final chapter in a story that's spanned centuries, ...This memorial has been created by the people of Great Yarmouth as all those pound coins have come together to make this happen. ... It's like the suspension bridge story has finally been told and the people who died have finally been the given the respect they deserve.[2]

The memorials themselves are testimony that, in the 21st century, the memory of these disasters is still important to their respective communities and they wish future generations to remember the victims.

Although subscriptions were sometimes raised for this purpose[3], in the Victorian period, physical memorials to disasters are few and far between, and even fewer list the victims by name. In Yarmouth the fund-raising effort was to concentrate, as Reverend Mackenzie put it, on "the yet living" rather than the dead. That did not mean that the people of Yarmouth were to forget the disaster. Elegiac poems were written[4] and memorabilia was readily available to tourists visiting the town[5]. Indeed, the following year Charles Barber, the Yarmouth bookseller, and member of the coroner's jury, published his "*Guide to Great Yarmouth*" for tourists to the town; Robert Brooke Utting, the brother of one of the victims, Sarah Utting, illustrating the book. The sentiments in this small book tells us a little about how they might have wanted the disaster to be remembered:

The 21st century Yarmouth suspension bridge disaster memorial.

Who shall describe the fearful impressions upon the minds of the spectators on the shore! Who shall paint the various passions and emotions which exhibited themselves in their countenances! Who shall tell of the anxiety of mothers who looked for their offspring — of husbands who sought their wives — of wives who cast a wild look of enquiry for their husbands — of brothers, sisters, and friends, who sought those who were dear to them, to be assured that they were not among the number engulphed!

What pen, what pencil, could convey a sufficient idea of the fearful tumult that ensued in the endeavour to escape the gurgling waters, and in the efforts of the bystanders to save the drowning! Those who witnessed that fearful scene will never become forgetful of its features — those who saw it not, can never fully feel its horrors. Many were rescued, and many were restored from almost the gates of death — and yet seventy-nine out of the number of those who went out to witness the mummery, never told its tale!

From an event so sad, so awful, we turn away to pursue the plan of our Guide Book; and yet we would not that the reader should too speedily dismiss from his mind those contemplations, which it is one of the great designs of an all-directing, and an infallibility wise providence to produce. Of these seventy-nine it naturally strikes us — "how engrossed with worldly occupations — how thoughtless of any others —how secure apparently of life — and yet how suddenly plunged into eternity!"[6]

C. J. W. Winter's representation of the disaster.

Another poignant reminder was by Cornelius Jansen Walter Winter, a reputable artist in the town, who produced an oil painting in 1846 depicting the collapse. Inscribed on the reverse is the following, "Falling of the suspension bridge Gt Yarmouth on Friday evening May 2nd 1845. & upwards of 100 persons drowned" and "desd & painted by C.J.W. Winter an eye witness of the occurrence"[7]. He also sold lithograph versions of his painting in his Yarmouth studio.

In more recent years, the Great Yarmouth Local History and Archaeological Society unveiled a blue plaque to the disaster in 2008 near the suspension bridge site. This was stolen in 2011 and a replacement was put on the Swan Inn nearby.

LEGACY AND MEMORY

In legacy terms, one can suggest that the Yarmouth bridge disaster contributed to the eventual process of introducing regulations governing infrastructure projects. Reactions to the collapse can be seen in the context of increasing clamours for state intervention to protect the public:

> We should be extremely sorry to prejudge the cause of the misfortune which must so soon become the subject of legal inquiry. The matter is still open, no doubt, to the proof, for aught we know to the contrary, that it was the result of unavoidable causes not within the control, nor necessarily, within the cognisance or calculations of the architect, the engineer or the contractor, or of the surveyor, or committee who may have approved and reported upon the design and execution of the bridge. But let us take the fact to be so; and even on the most favourable assumption how powerful a case is made out for the establishment of some responsible jurisdiction — some national commission of scientific and practical men, who could be invested with the important trust of minutely examining and severely testing every work of this kind before the proprietary or direction shall be at liberty to open it to the public.[8]

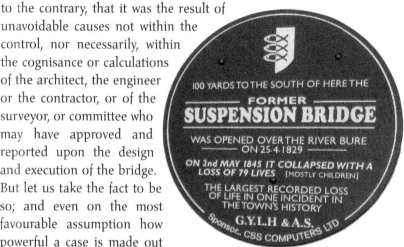

The plaque erected by the Great Yarmouth Local History and Archaeological Society recalling the disaster.

and after the publication of Walker's report, *The Morning Chronicle* stated:

> What a lamentable sacrifice of human life has been here made by a want of precaution, which would disgrace the most barbarous and improvident race! Our old system of decentralization of Government taking charge of, and looking to nothing, was advantageous — perhaps was possible — before we pressed into our common service the greatest and most powerful agents of nature. When our bridges were of solid masonry, and of no hazardous span; when we travelled at most a dozen in one vehicle; when pillars and rafters, the necessary strength and solidity of which were traditional proportions known to the commonest carpenter, were alone

employed; and when, in fact the risk to life was more a thing of affecting individuals than masses, then perhaps, we might leave men and things to take care of themselves. But now that we can do nothing which is not in the aggregate — when we transport, not by dozens, but by thousands — when we bridge over seas, and roof over acres, with thin rods of iron, which economy renders as slight as calculation declares to be safe — when by sea and land the lives of the hurried, the unwary, the helpless multitude are at the mercy of the machinist, and are still more periled each day by the multiplicity of engines and their age — the country is really called upon to modify its old rule of letting everyone care for himself. ...

Indeed there is no necessity for arguing the point. It is too self-evident, And we are applying the remedy here and there, by piecemeal, without unity or system or effect. ... Here alone we are too jealous of public liberties to permit the public safety to be cared for. And yet we give up the surveillance of streets and of our persons to a numerous and arbitrary police, and in a hundred ways we permit ourselves to be shackled and annoyed by the foreign system of tutelage, without the real benefits of that system.

The want of inspection, of authority, and a fitting tribunal in these great public concerns is beginning to be sorely felt. Had such a direct power existed ten years since, the quality of capital and labour which would have been economized in the mere laying out of railroads is incalculable. And the love of liberty and hatred of centralization which have forced us to dispense with such power, and which left the public and competing companies to their rival efforts, have wasted more money, cut up more estates, and given rise to more arbitrary wrongs, than a central board, however liable to dictation and misrule.

But indeed the experience of the session — the utter inability of either Parliamentary commission, or an improvised board without authority, to decide between conflicting interests, renders a charge in our administrative system, relative to public works, indispensible. A portion of the system should be the duty of providing for the public security in the execution of all those great works, on the solidity and fitness of which depends the lives of masses of the population.[9]

However, it was not protestations in newspapers that led to change, but the time it was taking to process the number of private bills though Parliament. In 1846 the government set up a Select Committee on private bills, "to examine the applications for local acts during this session of Parliament". Their specific brief was to concentrate on those bills that were concerned with the improvement in public health, but the Committee concluded that their recommendations could be applied to all private bills. Rather than pass individual acts, general acts should be used, and the responsible government department, such as the Home Office, the Board of Trade, the Admiralty, the Commissioners of Woods and Forests, the Enclosure Commissioners, or others, as the case may be, should be given the power to sanction them. As far as bridges were concerned, the Preliminary Inquiries Acts of 1848 and 1851 required those who sought, "to construct any works on the shore of the sea, or of any creek, bay, arm of the sea, or navigable river communicating therewith, or to construct any bridge, viaduct, or other work across any creek, bay, arm of the sea, or navigable river, or to construct any work affecting the navigation of any harbour, port, tidal water, or navigable river" to submit their plans to the Admiralty, and gave the Admiralty Commissioners the power "to appoint a competent person or persons to be an inspector or inspectors, for the purpose of inquiring, in such manner and at such time and place as they shall direct, into all such matters as they shall deem necessary to enable them to report to Parliament their opinion upon every such Bill". In 1862, these powers were transferred to the Board of Trade. Although this did not mean the type of scrutiny that was being called for after the Yarmouth disaster, it did mark a shift towards more state regulation.

Within the town itself, the removal of the bridge enabled the town to grow and take advantage of the economic opportunities offered in the latter half of the 19th century, such as, the significant growth of its herring industry and its development as a tourist resort.

The Yarmouth Suspension Bridge disaster of 1845 not only provides an interesting insight into a small coastal town in the 1840s, but also the wider context of 19th century Britain. Today we talk about the speed of change, but it was no less true of communities in this period. Robert Cory Jun. believed he was building a bridge that would not only stand the test of time but also be his legacy for generations to come. However, by 1844, his masterpiece proved hopelessly inadequate for the economic upheaval that was about to overwhelm the town. Cory's turnpike road to Acle across the marshes was the last of its kind in Norfolk, while the Yarmouth and Norwich Railway was the first. A mere twelve years between them, what was thought to be a huge advance had so soon been superseded.

Initially railways had been about moving raw materials to the coast, Yarmouth's railway was about moving produce inland and to centres of population. Its economic significance cannot therefore be underestimated[10], nor can the cost, which put Cory's investment into the shade. No longer could one individual provide the investment needed for the economic advancement of the town.

The mackerel fishery at Great Yarmouth.

The dispute over the bridge in 1844 signified a turning point. The civic elite of this ancient borough had always been its merchant class mixed with those in the professions (clergy, law and medicine). While this remained the case, their control was to slowly diminish in the following decades, whether it was to external commercial interests, or to central government. When the railway opened in 1844, only one Director of the railway company was a Yarmouth man, and that was the banker and brewer, Edmund Lacon. The fears expressed over the 1844 bill shows that this elite realised that commercial and municipal affairs were not necessarily in their control. The collapse of the bridge highlighted this paradox, a fact that was not lost on the people at the time. The *Bury and Norwich Post* looked forward to the railway company being integrated into the commercial and political fabric of the town:

> We are exceedingly glad that the differences which hitherto have existed between the proprietors of the bridge and the directors of the railway have been brought to an amicable conclusion, and shall be equally delighted to see the company assume that position with the town of Yarmouth which we have ever been of the opinion it might have held. There is evidently a greater disposition on the part of the Company to consult the interests and convenience of the inhabitants of this important town, and we shall

hail with satisfaction the manifestation of an accommodating disposition by the local authorities.[11]

It was a new state of affairs and everyone had to adapt; as did the individuals affected by the disaster. For example, the Livingstones who lived above their draper's shop on the corner of the Market Place and King Street had other children and eventually, when William died, William junior, who had survived the disaster, took over the business.

Charles Cory, the proprietor of the bridge, became Town Clerk in 1851 and continued in that office until his death in June 1869. The Town Council at a special meeting convened on 21 June passed the following resolution:

> That this Council desire to record their deep sense of the loss which they have sustained by the untimely death, on the 10th of June inst., in the 57th year of his age, of Charles Cory, Esq., which sad event took place at Lugano, in Switzerland, after a brief illness. From the time of his appointment in 1851, Mr. Cory discharged all the duties of his office of Town Clerk with singular ability. The rights and privileges of this Corporation were carefully watched and successfully defended by him, and our records bear ample testimony that during his period of office he conducted to completion many important works of public utility which will confer a lasting benefit on the town. Both as a public servant and a private gentleman his death is much to be deplored, and the Council are unanimous in their desire to pay this tribute to his worth.[12]

Following his death, a memorial window was dedicated to him in the south aisle of St. Nicholas Church following public subscription[13].

After his departure from Yarmouth, Arthur Nelson was to remain with Cooke's Circus for its ten week run at the New Royal Standard in Shoreditch, but there is no mention in the press of the clown performing for the rest of 1845. Perhaps his association with the collapse had been damaging not only to his mind and body, but to his reputation as well. At the beginning of 1846 he went to America, and, away from any limelight that might have shone on his antic, he managed to resurrect his career. He returned to Britain for the winter season of 1846 and, engaged by William Cooke, he certainly performed in Norwich in September[14], and may well have done so in Yarmouth during their tour. By 1850, Arthur Nelson was describing himself as the "Clown King[15]", a title he was to use for the remainder of his life. Throughout 1853 he was performing with Madame Warton and her troupe of tableau vivant or pose plastique[16]. This form of amusement had become popular in the 1840s. Actors, mostly women, would impersonate renowned paintings or classical statues clad in "fleshings". A similar technique had also been employed in early performances of Thomas Cooke's circus to recreate heroic scenes from the past before the action started. Following her death in February 1854, Nelson took over the management of the troupe[17] and by August 1855 was also managing Mr. E.T. Smith's Royal Circus (formerly Hernandez & Stone), as well as taking his 230th sail on water in a tub drawn by 4 geese in Maidstone[18]. The increasing number of advertisements[19] he placed in the Era in 1856 may show that engagements were difficult to find during that year, but things picked up in 1857 with the high point being the organisation of the "AQUATIC SPORTS, including a GRAND CARNIVAL AND WATER PAGEANT" at Peel Park, Bradford in June 1857. On this occasion he did his favourite stunt for the 334th time[20].

Nelson died on 27 July 1860. His obituary in the Era read:

Death of Mr. Arthur Nelson, the Clown.

The Death of Mr Arthur Marsh Nelson, popularly known as the "clown king," took place at Burnley, on the 27th ultimo, and he was interred at the Burnley Cemetery a few days afterwards. He was born about the year 1811, and on his first introduction to the stage played the leading parts in the legitimate drama in provincial and minor theatres. He subsequently adopted the "talking clown" as his vocation, and his repartees in the ring were often remarkable for their readiness and humour. He was a good musician, and his clever performances of the pine-sticks exhibited skill and, with which, he had cultivated the eccentric branch of the art he had adopted. At Vauxhall Gardens and at several of the minor metropolitan and provincial theatres he was a great favourite. His last appearance in London was at the Alhambra Palace a few weeks back, when he introduced the female horse-tamer to the public.[21]

Arthur Nelson's memorial stone in Burnley cemetery.

At their deaths, neither Charles Cory nor Arthur Nelson was remembered for his role in the disaster of May 1845 — Charles Cory was remembered for his dedication and loyalty to the people of Yarmouth; Arthur Nelson, for his musical talent and humour in the ring, rather than his act with geese and a washing tub.

Today, our motives for remembering a historical event are often complex. Julie Staff was moved by the victims and felt there were few visible signs of this event to be seen in Yarmouth. Kevin Abbey, who carved the 2013 memorial, was motivated by the history and that the community should remember those that died[22].

In 2001 the Department of Culture, Media and Sport defined memorials as "stories in stone", heritage objects that serve, "as a means of stimulating and conserving memories transmitted from generation to generation"[23]. This can only be the hope of all those that are interested in remembering the past and this disaster in particular.

[1] *The Scotsman website* — 27 December 2013, available at http://www.scotsman.com/lifestyle/heritage/tay-rail-bridge-disaster-victims-to-be-honoured-1-3248610 (last accessed 17 March 2015).

[2] *BBC News website* — 28 September 2013, available at http://www.bbc.co.uk/news/uk-england-norfolk-24240357 (last accessed 17 March 2015).

[3] An example is the Rotherham or Masbrough Boatyard disaster of 1841. On 5 July the vessel "John and William" was to be launched from a boatyard at Masbrough with over 150 men and boys on-board while on the slipway. As it slid sideways into the canal, in similar circumstances to the Yarmouth disaster, those on board rushed to the leeward side of the vessel to see the effect of it hitting the water. The boat overturned throwing them into the water and trapping them under the hull. Over 50 died, 47 of which were under 21. The memorial in Rotherham Minster names the victims and their occupations. See http://www.rotherhamweb.co.uk/h/botdis.htm (last accessed 16 March 2015).

[4] Printed versions of two of these poems survive, "Consolatory Lines on the Downfal of the Suspension Bridge at Great Yarmouth On Friday, the 2nd Day of May, 1845; Addressed to the relatives and friends of those who were lost. By Mrs. Hart." and "The Parting and the Meeting Or The Burial of Yarmouth Bridge, James Stuart Vaughan, 1847".

[5] *Norfolk News* — 27 September 1845, p.1. — Charles Barber, a book and printseller in the town, and one of the jury members, had purchased the engraving used by the Norfolk News in May and was selling prints to "the Trade in very reduced Terms. Sold Retail, One Penny each; or on best drawing paper, 2d.".

[6] Anon. (1846) *Guide to Great Yarmouth; with thirty four illustrations by Brooke Utting* p.38-39 available at https://books.google.co.uk/books?id=OJJeAAAAcAAJ&dq (last accessed 19 March 2015).

[7] This painting can now be seen in the Tide and Tide Museum in Great Yarmouth.

[8] *The Times* — 7 May 1845.

[9] *Morning Chronicle* — 28 May 1845, p.4.

[10] Robert Stephenson, *Inaugural Address to the Institute of Engineers*, January 1856— "In the fish trade indeed railways have caused and are causing a prodigious revolution. Large fishing establishments have been formed at different parts of the East Coast. Before the Norfolk Railway was constructed, the conveyance of fish from Yarmouth to London was entirely conducted in light vans with post horses, and was represented by a bulk of about 2000 tons a year. At present, 2000 tons of fish are not unfrequently carried on the Norfolk Railway, not in a year but a fortnight." cited in Nail, J. G. (1865) *Chapters on East Anglian*

Coast. Part 1. Great Yarmouth and Lowestoft; Their Topography, Archaeology, Natural History, &c., and a History of their Fisheries. p.343.

11 Bury and Norwich Post — 25 June 1845, p.3.

12 Bury and Norwich Post — 22 June 1869, p.7.

13 Palmer, C. J. (1872) *The Perlustration of Great Yarmouth with Gorleston and SouthTown* p.54 available at https://archive.org/stream/perlustrationofg01palm#page/54/ (last accessed 19 March 2015).

14 Norfolk News —12 September 1846, p.3.

15 Lloyd's Weekly Newspaper — 31 March 1850, p.10.

16 The Era — 21 August 1853 p.1. — "THE ORIGINAL AND CELEBRATED MADAME WARTON and TROUPE from the WALHALLA, LEICESTER-SQUARE, LONDON are now making one of the most successful tours ever known. In every town they visit the audiences are numerous and highly respectable. The Magistrates and Clergy are unanimous in their approval of the Performances. Nelson, the Clown King, accompanies the Troupe; his inimitable performances on the Pine Sticks and Rock Harmonicon, give great satisfaction. The Public are requested to notice the spelling of the name WARTON.

N.B. — Madame and Troupe will shortly be at liberty to receive engagements with London Managers.".

17 The Era — 27 August 1854, p.1.

18 The Era — 19 August 1855, p.11.

19 The Era — 7 June 1857, p.1. is one example, "Mr. ARTHUR NELSON, CLOWN KING and CHAMPION MOMUS OF EUROPE, the only clown that ever made 25,000 Persons laugh all at one time (*vide* Bradford Press) will be at liberty to engage with Circus and Theatrical Managers, on Monday June 8." — Nelson's reference is probably to the Gala at Peel Park, see note 20.

20 Bradford Observer — 21 May 1857, p.1.

21 The Era — 26 August 1860, p.10.

22 BBC website — 28 September 2013 — "To be so involved in a big part of Great Yarmouth's history means a lot. It's important to have a lasting memorial for those who tragically lost their lives on what should have been a fun occasion and for the town to be able to remember these people at a memorial where they can come and pay their respects." available at: http://www.bbc.co.uk/news/uk-england-norfolk-24240357 (last accessed 23 March 2015).

23 *The Historic Environment: A Force for Our Future* — Department of Culture, Media and Sport 2001, available at http://webarchive.nationalarchives.gov.uk/+/http://culture.gov.uk/images/publications/historic_environment_review_part1.pdf (last accessed 17 March 2015) cited in the article *A Story in Stone: the Tirah War Memorial in Dorchester*, Martin D., The Historian number 123 Autumn 2014, available via subscription at http://www.history.org.uk/file_download.php?ts=1416394951&id=15045 (last accessed 17 March 2015).

Appendix 1

Narrative of the Commencement and completion of the Suspension Bridge over the River Bure at Great Yarmouth erected by Mr. Robert Cory jun. in 1828 and 1829. This is a transcript from Robert Cory Jun unpublished manuscript bound in a volume called, "The Suspension Bridge" and dated 1832.

The title page for Cory's manuscript

In 1810 I purchased an Estate at Runham opposite Yarmouth North Quay comprising a Public house and Gardens called Vauxhall of about 100 acres of land together with a ferry and tolls over the River Bure.

Long before I purchased the Estate I had entertained an idea that at no distant period a Bridge would be made across the River hereabouts and a Turnpike Road from Yarmouth to Halvergate and so five years after about 1823 I projected such a Road and went to a considerable expense for plans and estimates, gave notice of an application to Parliament and formed a committee of Gentlemen to carry it into effect but it was thought the Tolls would be inadequate to the expense and the measure was abandoned.

Afterwards several Gentlemen and solicitors undertook the same plans, knowing it must materially benefit my Estate, I rendered them every assistance, however they are proved abortive.

One Morning in August 1826 coming from Horsey, I observed several sticks set up at various distances with papers on them on the Caister Road about a quarter of a mile from the Gates and on enquiry I learnt that a project was on goal for making a turnpike & building a Bridge across the

River near the Cinder boons, and that these sticks were the surveyors marks. This project would have avoided my land and made my Tolls valueless. Without disclosing my intentions, I immediately gave notices for an application to Parliament for an Act to build a bridge at my ferry, and before it was generally known, I obtained an Act which received the Royal assent 20 May 1827.

By this measure, and having a clause inserted in my Act, that no person should go over the River by any other medium within the limits of my ferry (which extends more than a mile up the River), their schemes were frustrated.

As soon as the Act was passed, I advertised for plans, received several, some stone, some for cast iron, others for brick and wood, and others for iron suspension over. I had often seen Brighton pier, and had some inclination for a chain bridge, and in October 1827, I went to look at one at Witney which had been built by Mr. Gibson. Mr. Goddard accompanied me and took the measurements, and soon afterwards I went to Brighton and on my way looked at the Hammersmith one. I then saw a small one at Ryde in Sussex, and afterwards, hearing there was one erected at Leeds upon a different principle, using rods being suspended from an arch thrown over the River instead of from chains over pillars. In January 1828 I went and looked at it, but did not like it, and determined upon one suspended from chains.

Mr. Goddard made plans and models for a suspension bridge, and I began to prepare the contract. In the meantime, my eldest son Isaac came to Yarmouth, and looking at the models, and upon Mathematical principles, a bridge built upon these models would not stand an hour, and he strongly advised me to submit the whole to some reputable architect, and recommended Mr. Scoles, a gent with whom he had travelled thro' part of Italy and Germany. To him therefore, I submitted Goddard's plans and models, and the bridge was built under his superintendance.

I then issued proposals for a contract which Mr. Goddard obtained for ¬£1825 appointed Mr. John Green surveyor of the works & accepted the contract on the 2nd June 1828.

On the 2nd May, some labourers were employed to ascertain what foundations might be expected when it was found necessary to drive piles.

On Monday the 9th June, I fixed the site and the excavation was begun on the West side of the River and carried down 16 feet below the surface.

On Saturday 17th June the first pile was driven and a good foundation attained there.

On the 10th of July (being Mrs. Cory's birthday & the children at home for the holidays) I determined to lay the first stone, but when the day arrived the piles were not in sufficient forwardness & consequently it was postponed.

A large stone with a hollow to receive a bottle was prepared & a few coins of Geo. IV with the following inscription to parchment were deposited in a bottle hormotically [sic] sealed :

The parchment.

On Monday 28th July 1828 everything being finished, I deposited the bottle in the Stone & as the clock struck twelve, I laid the first stone in the presence of my wife, my brother, my children Marian, Harriot, William, Charles, Robert, Alexander, Laura, & John (Isaac, Horace & Louisa being in London). My friends Miss Preston, Miss Marian Preston & Miss Charlotte Preston, the three daughters of Mr. John Preston, Mrs. Scott, The Rev. John Scott, My old friend, Mr. John Preston, the Rev. Wm. G. C. Courtalay, Mr. Daniel Preston, Mr. Francis Preston, Miss Sylet, Mr. Green, Mr. Goddard & his son Henry, each of whom laid a brick. 100 then had a luncheon on the Green. I regaled the workmen with bread & cheese & beer & the Prestons came & spent the morning at home.

The list of those present.

On the 11th September, the brick work was finished on the West side of the water & the foundations began on the East; here we had a small amount of trouble, the ground having then made [built up] but a few years from the muck holes where the muck of iron then had been deposited for centuries, the piling was obliged to be very thick & deep, a very considerable extra expense was incurred.

The spaces between the works of the bridge & the excavations were filled up with mud from the River, which the Haven River Commissioners kindly gave me & which took upwards of 2500 tons.

On the 22nd October Mrs. Cory & myself went to see the last stone laid, which was on the North Side of the East Bank of the River. This Stone weighed more than two tons & just so it was raised to its proper height and, as a man was getting under it to spread the mortar, the Rope broke & the Stone falling upon the edge knocked down part of the pier & crushed the brickwork of the wall, but luckily no one was hurt and as little mischief done as possible, in three days after everything was restored & the stone deposited in its proper place.

The piers & Columns being completed on Thursday the 19th February 1829 the Suspension Chains were thrown across the River; they were raised by means of a platform laid upon the top of the new hand Engine (which the Commissioners lent us). The Engine was moved across the River, and the whole were suspended in the most masterly manner in two hours & a half. After the Chains were suspended, I set the place for the tolhouse.

On the 26th February, the Suspension rails were all hung & the Bearers for the platforms fixed by means of a temporary plank. I first walked over on the following day. Mrs. Cory and Louisa walked over on the first of March with the Miss Prestons; these were the first persons who passed over, as I prohibited all others.

After this the Bridge went easily on to its completion. My Arms & Crest were put up on the Column & the following inscriptions were cast on the iron plates on the Base on the North side of the South pedestal.

```
PONTEM · HVNCCE
D · S · P ·
FACIVNDVM · CVRAVIT
R · CORY · R · FILIVS
ANNO · SACRO
M · DCCC · XXIX ·
J · I · SCOLES · ARCHITECT ·
```

North side inscription.

On the South side of the North Pedestal

> THIS BRIDGE
> WAS BUILT BY
> ROBERT CORY JUN.
> INSTEAD OF AN
> ANTIENT FERRY
> A·D· MDCCCXXIX·
> G. GODDARD CONTRACTOR

South side inscription.

On the 23rd of April 1829 (being the King's birthday kept), I determined to open the bridge & for that purpose made arrangements with John Preston Esq., the Mayor, to have a procession from the church. I then issued cards of invitation to the Corporation, Clergy & principal inhabitants & their sons & Mrs. Cory invited about forty ladies. I issued cards also for a Dinner Party & a Ball in the Evening at the New Hall concluded the day.

On the morning of the 23rd April, the day was ushered in by the Ringing of Bells, but the rain fell in torrents so that I was afraid the procession would be spoiled, however, we set the tables for the lunch along side of the House under cover instead & another table in the large room for the ladies.

About eleven o'clock, the rain ceased & the Corporation went to Church. At twelve o'clock they assembled at the Guildhall where they were met by about a quarter past 12 they proceeded in the following order:

The Bellman
Beadles
Constables
Corporation of Flags
Band of Music

The Regalia
Mayor & Recorder
Substeward & Minister
The Aldermen in Scarlet gowns
The Clergy in their Robes
The Common Councilmen in Gowns
The Gentlemen
The Hospital, Charity and Workhouse children

The procession went from the Guildhall over Fullers Hill & over the Bridge. When the Mayor arrived on the centre of the bridge, eight guns were fixed. The Quay was crowded & { } was very imposing.

At the Lunch on the Green, the King's health was drank with 3 times 3, and the children, about 4 or 5 hundred, sang God save the King accompanied by the Band. The Mayor then gave my health & { } to the Bridge was drunk with three times three to which I returned thanks after the luncheon. The procession paraded round the Green in the same order & returned to the New Hall where they unrobed. I went to the Ladies & drank wine with them. At about 1/2 past 2 all left the Green.

The workmen I gave a dinner to, on the Green & sent each of the people in the Fisherman's Hospital (to which I was Treasurer) a quart of porter and a 6d. loaf.

Eighteen Gentlemen afterwards dined at my House & in the evening we went to the Ball which was well attended.

On coming from the Bridge I picked up a bent sixpence in the Road which was looked upon as a lucky omen.

It is worthy to be remarked that, not withstanding the number of workmen & length of time they were employed, not a single accident happened to the injury of anyone.

The Tolhouse was built, after a design by Mr. Scoles, by Mr. Gooda [Goodard], the Contractor.

A few weeks after I was presented with a silver medal & each of my children with a copper one by William Barth Esq., which was struck by a subscription of a few Gentleman.

There was only one silver & 50 Copper struck. I had the silver & 12 Copper; Mr. Barth had 12 & the other subscribers one or two each - It cost 25 gs.. I was told about 6 were sold at 7/6 each. They are now rare.

In the following year I set myself heavily to get a Turnpike Road from my Bridge to Acle which notwithstanding many opponents, I fully succeeded

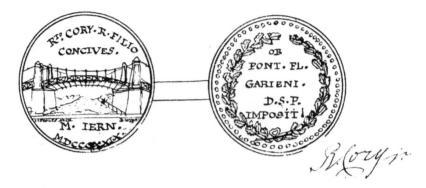

The reverse sides of the medals struck to commemorate the opening of the bridge.

in "an Act for making and maintaining a Turnpike Road from the Bridge over the River Bure at Great Yarmouth to Acle (with certain Branches therefrom) all in the County of Norfolk" received the Royal Assent, the 3rd May 1830 & the Road was immediately after begun.

The overseers of Yarmouth having laid me heavily to the poor Rates & the inconvenience of pulling Horses up to the Toll on the descent of the Bridge being great, I determined to build a Tollhouse on the Runham side which I did. It was planned by Wilson John (who is with Mr Scoles the Architect a pupil) & was begun in April 1837 and completed in Sept. following - To this I attached a piece of ground as a garden - John Fountain who I first appointed collector of the tolls died in March 1837 and I appd [appointed] Charles Bailey who took possession 25 March & went into the new house, & began to collect tolls there 29th Sept.

The former Tollhouse I now let for ¬£5 a year.

The cost of the new Tolhouse Garden etc. was ¬£ [sic]

Appendix 2 – list of victims

	Surname	Forename	Victim's age
1	Adams	Robert	7
2	Auger	Caroline	10
3	Balls	Reeder Thurston	16
4	Barber	Christopher	11
5	Beckett	Ann	8
6	Beloe	George John Henry	9
7	Borking	Emily Hansworth	5
8	Bradberry	Isaac	20
9	Buck	James Seaman	4
10	Burton	Benjamin Patterson	7
11	Bussey	Harriott	26
12	Buttifant	Sarah Ann	18
13	Church	Caroline	16
14	Church	James	7
15	Cole	Jane	16
16	Conyers	Elizabeth	13
17	Crowe	Eliza	14
18	Ditcham	Mary Ann	64
19	Duffield	Eliza	10
20	Durrant	William	12
21	Dye	Charles	2
22	Dye	Benjamin	9
23	Ebbage	David	9
24	Edwards	Maria	12
25	Field	Hannah	12
26	Field	Susannah	13
27	Fox	John Horace	19

28	Fulcher	James	14
29	Fulcher	Elizabeth	16
30	Gilbert	Sarah	12
31	Gotts	Alice	52
32	Gotts	Alice	9
33	Grimmer	William	8
34	Hatch	Elizabeth	11
35	Hendle	William	10
36	Hunn	Sarah	13
37	Hunnibal	Elizabeth Jane	12
38	Jenkerson	Mary Ann	10
39	Johnson	Thomas (or Robert)	8
40	Johnson	Elizabeth	8
41	Johnson	Sarah Ann	16
42	Juniper	Maud	9
43	King	Mary Ann	11
44	Lake	Mary Ann	2
45	Little	Harriet Mary	13
46	Livingstone	Joseph	6
47	Livingstone	Matilda	7
48	Lucas	Frederick	62
49	Lyons	William	6
50	Manship	Elizabeth	28
51	May	Clara	20
52	Morse	Robert	26
53	Mears	Susan	8
54	Morgan	Elizabeth	62
55	Parker	Charlotte	8
56	Powley	Elizabeth	21
57	Powley	Richard	4
58	Powley	Amelia	10
59	Read	Elizabeth	5
60	Richardson	Phoebe	17

61	Roberts	Lydia	12
62	Roberts	Mary Ann	19
63	Scotten	Ann Maria	20
64	Stolworthy	Maria	14
65	Tann	Harriett	15
66	Tennant	John	11
67	Tennant	William	10
68	Thompson	Mary Ann	15
69	Thorpe	Heppy	12
70	Trory	William Townshend	12
71	Utting	Sarah	18
72	Utting	Louisa	7
73	Utting	Caroline	9
74	Vincent	Maria	19
75	Vincent	Richard	unknown
76	Watts	William Walter	9
77	Yallop	Martha	20
78	Young	Emily	6

78 individuals lost their lives on 2 May 1845. Those on the "missing list" had all been identified by the evening of 3 May apart from James Seaman Buck and Louisa Utting. Their bodies were not found and identified until 17 and 27 May respectively.

The jury refrained from giving verdicts on the deceased until they had agreed the cause of the disaster and therefore it was not until 2 June, that the Coroner took their verdict on Louisa Utting, given on 27 May, to form the basis of verdicts on all those who could be proved to have been on the bridge at the time:

> That the deceased (Louisa Utting,) came by her death by the falling of the suspension bridge across the River Bure in this borough, on the 2nd of May, 1845. That the falling of the bridge was attributable immediately to the defect in the joint or welding of the bar which first gave way, and to the quality of part of the iron and workmanship being inferior to the requirements of the original contract, which provided that such should be of the first quality.[1]

We are told by the *Norfolk Chronicle* that, "in seven or eight instances where it was found impossible to trace them on to the bridge, and on these a verdict of found drowned was recorded."[2] 77 verdicts were recorded that day, and with the previous verdict on Louisa Utting, the total remained 78.

Despite this evidence, throughout May and June 1845 many provincial newspapers up and down the country[3] and other contemporary documents stated that 79 was the final number. As a result this figure has invariably been used to account for the number of fatalities even up to the present day[4]. One cannot be sure where the number 79 came from. However, the Norfolk News on 17 May 1845 emphatically stated, "The entire number drowned is 79, and no more" and this account may be the source of the discrepancy. The syndication of text among national and provincial newspapers meant that that number was circulated widely[5].

[1] *Norfolk Chronicle* — 24 May 1845, p.3.
[2] Ibid. — 7 June 1845, p.3.
[3] For example, *Preston Chronicle* — 17 May 1845 p.4. — "It has been ascertained that the loss of life by the fall of Yarmouth suspension bridge was overstated; it amounts to 79 persons."
[4] The most likely candidate for perpetuating this number was Charles Barber, a local printer and member of the jury. From late June he sold prints of the illustration of the disaster he had purchased from the *Norfolk News* and added a caption "as it appeared after the accident by which 79 lives were lost, on the Friday evening, May 2, 1845."
[5] The most common text used by provincial papers, even as late as June 1845 was, "Great Yarmouth. — The total loss of life arising from the accident at Yarmouth is 79 (bodies found 77), and most providentially not one of that number had any dependent on him or her. Only eight or ten of the deceased exceeded twenty one years of age."

Appendix 3

Miscellanies relating to Great Yarmouth

This is a transcript from Robert Cory Jun unpublished manuscript, "Miscellanies relating to Great Yarmouth" by R. Cory with additional notes by J. Davey dated 1846. It is included because in this instance it gives the dimensions of the bridge.

177
Suspension Bridge

In 1827 Robert Cory Jnr Esq (who was owner of the public gardens and land called Vauxhall Gardens, and the Ferry Farm estate opposite the North Quay) applied for and obtained "An Act of Parliament for erecting a Bridge over the River Bure from Runham to Great Yarmouth in the County of Norfolk", to which the Royal Assent was given 28 May 1827. In 1828 contracts were advertised for building a Suspension Bridge upon the plans and models of John Joseph Scoles Esq Architect, and Mr Godfrey Goddard Whitesmith entered into a contract with Mr Cory dated 2 June 1828 for £1825. After signing the contract it was found necessary to pile the foundations and on the 8th June the excavations were begun on the west side of the river and the first pile was driven in the 17th June.

On the 10th July (being Mrs Cory's birthday) it was intended to lay the first stone but on account of the piling it was deferred to the 28th July when it was laid by Mr Cory in the presence of his family and friends and the following inscription on parchment was, with a few coins of Geo 4th put into a bottle and deposited in a cavity in the stone.

"This Bridge was erected by Robert Cory the Younger of Great Yarmouth in the County of Norfolk Esq. to the north and in lieu of an Ancient Ferry belonging to him under the Authority of An Act of Parliament passed in the 8th year of the Reign of his present Majesty King George the Forth
-and
This the first stone was laid by the said Robert Cory in the presence of his family and friends some of whose names are subscribed the 28th day of July in the year of our Lord 1828 In the Mayoralty of John Mortlock Lacon Esq Mayor of the Burgh of Great Yarmouth May God prosper the undertaking"

The last stone was laid on the 22nd October which weighed more than two tons.

On the 19th February 1829 the piers and columns ~~being~~ were completed and the Suspension Chains were thrown across the river by means of a platform laid upon the top of the mud engine in the short space of two hours and a half.

On the 27th of February Robert and Mrs Cory walked over the bridge on a plank laid on the bearers and on the 22nd April 1829, being the day in which the King's birthday was kept the Bridge was opened and after the Church the Mayor, Recorder and Corporation the Clergy and upwards of 100 gentlemen assembled at the Guildhall and proceeded in the following order

Bellman
Beadles
Corporation flags
Band of Music
The Regalia
Mayor and Recorder
Substeward and Aldermen in Scarlet robes
The Clergy in their robes
The Common Council in Gowns
The principal inhabitants
Charity, Hospital and Workhouse Children

On arriving at the Bridge they were received by the workmen and eight guns were fired. The North Quay was crowded with spectators. After partaking of a lunch on the Bowling Green the procession returned and went to the town hall and unrobed.

The Tolhouse on an elegant plan by Mr Scoles was also completed at the same time and Mr John Fountain was appointed to keep it.

The following inscription is cast on the iron plate on the north side of the South Pediment

PONTEM . HVNCCE
D.S.P.
FACIVNDVM . CVRAVIT
R . CORY .R. FILIVS
ANNO . SACRO
M . DCC. XXIX
I . I. SCOLES. ARCHITECT
and on the opposite Column

"This Bridge was built by Robert Cory Junr instead of an Ancient Ferry AD 1829 by G. Goddard Contractor."

On the Columns are the arms and crest of the proprietor.

Soon afterwards a few gentlemen of the town subscribed for a Medal and presented Mr Cory with one in Silver for himself and a bronze one for each of his children which were presented by William Barth Esq

Dimensions sic.
Span of Arch 86 feet
Length of Bridge from foot to foot 186 feet
Breadth of footpaths 4 feet each
Breath of railway 6 feet 9 inches
Height of Columns from Pediment 14 feet
Height of Pediment from Quay 7 feet 4 inches
Weight of Iron Chains each 52c 2q 3lb
Weight of Iron 18 tons 6 cwt 3gss 27lb
Dragstones 264 Cubic feet
Stone 1354 Cubic feet
Brickwork 1499 yards
Foundation timber
Piles 120 1057 feet
Plank 2410 feet
Bond timber 148 feet
Cost of Bridge
Act of Parliament 419.9.0
Contracts etc 58.9.0
Bridge 2425.16.0
Tollhouse 127.13.9
Approaches etc 49.8.6
Architect, Surveyors etc 150.18.6
Expenses laying first stone, opening bridge etc. 53.10.9
3285.5.6

In 1837 the overseers having rated the Bridge which not only burdened it with Poor Rates but Church rate Paving & Lighting etc accounting to nearly £20 a year I built a new tollhouse in the Runham side of the water about 100 yards from the foot of the bridge where the tolls are now collected. Charles Bailey Collector, and I let the tolls to him for £250 a year I also let the old Tollhouse for £5 per annum Notwithstanding this the overseers still rated me and I appealed which was determined as our Quarter Sessions and I got 40s Costs.

They still threaten to rate Bailey.
R.C.

On the 2nd of May 1845 a number of persons having assembled on this bridge to witness a Clown from Cookes' Equestrian Circus swim in a tub drawn by geese, one of the suspending chains gave way & precipitated the mass of people into the water, by which accident 79 lives were lost. It is very remarkable that of this large number not one head of family was removed so as to leave the survivors dependent on parochial relief. The scene was the most distressing I ever witnessed. J.D.

Index

Abbey, Kevin 103
Acle 10, 61, 64, 68, 74, 99, 111-112
Adam and Eve Garden 18
Arnold, Mary Ann 27
Astley, Philip 12, 77-78, 80-81, 89, 91
Astley's Amphitheatre 78

Bailey, Charles 61
Bale, Mr 20
Balls, Reeder Thurston 19, 27
Barber, Charles 94
Barlow, W. 86
Barry the clown 80
Bath Place 27
Becket, Ann 19
Beckwith, A.A.H. 67
Beloe family 30
    Arthur 31-33
    George Henry John 28-29, 34
    John 33
    Louisa 29
Bentham, Jeremy 10, 56
Bidder, George Parker 71
Borking, James 16, 113
Borking, Eliza Eliza 16, 113
Bradberry, Isaac 20
Breeze, Mr. 33
Breydon 14, 16, 66, 74
Brick Quay 72
Brighton pier 59, 106
Buck, James Seaman 19
Bure, River 6, 15-16, 43, 57-58, 60-61, 63, 65-70, 74, 76, 77, 105, 112, 116-117
burial club 23
Bussey, George 26
    Harriot 21, 26
Charlton the clown 83
Church Hall 19
Clerk, Town 44
Cole, Jane 24
Cooke, Alfred 83, 85-86
    George 86
    Kate 15
    William 15, 79, 82, 86, 87, 101
    Thomas Taplin 78
Cooke's Circus 15, 20, 78-80, 84-87, 89-90, 101-102
Coroner 19, 37, 43, 75, 116
Cory family 65, 67
Cory, A.T. 69-70
Cory, Charles 23-24, 34-35, 39, 43, 46, 67, 68-76, 99, 101-102
Cory, Isaac Preston 59, 69
Cory, Rev. R. William 69, 71
Cory, Robert 57
Cory, Robert Jnr (builder of bridge) 57, 59, 60-65, 79, 105, 107-110, 117-119
Cory, S.B. 63
Council, Great Yarmouth 45, 48, 63, 66-67, 74, 101, 118
crêpe 24, 30, 33

Dickens, Charles 23
Disraeli, Benjamin 9, 49
Dissenter's burial ground 24
District Visiting Society 50-51
Dowson, Benjamin 34, 50-51
Duff, Adam 70

Escott, Mr 45
Ferrier, Dr. W.S. Richard 19, 68
Ferry Boat Row 26
Field, Hannah 24
    William 24
Fountain, John 61
Freeman 17
Fuller's Hill 29, 33, 68

Gaunless Bridge, (West Auckland Bridge) 62
Ginnett's Circus 89
Gladstone, Rt. Hon. William 53
Goddard, Godfrey 59-60, 106, 108, 117, 119
Gotts, Alice 24, 26
    William 26
Graham, Sir James 45, 46
Greathead, Mr 26
Great Yarmouth Local History and Archaeological Society 96, 97
Grimmer, William 50
Grout Baylis's silk factory 24, 57

Halvergate 58, 105

Hammond, Richard 65
Harper's Magazine 21
Hart, Mrs. 37
Havel 8
Haven and Pier Commissioners 68, 74
Hills, Rev. G. 51

Illustrated London News 17-18, 20, 28, 35, 39, 40, 47, 55

Jetty Road 27
Johnson, William 68-69
John, Wilson 61
Jury 19, 44, 50, 75

Kelly, Fitzroy 72-73
King Street 101

Lacon, John Edward 71
Lacon, Sir Edmund Henry Knowles 63, 68, 71, 74, 100
laissez faire 11
last 8, 89
Laurence, John 70
Lilly, William 85
Limekiln Walk 70
Lime-Kiln wharf 65
Little, Harriet 24
Livingstone, Joseph 24
    Matilda 24
    William (son) 20, 101
    William (son) 20, 101
    Martha 20
Lound 21, 26, 27

Mackenzie, Rev. Henry 38, 47, 52-53, 56, 94
Market-place 31, 32, 68, 101
Milner, Henry M. 77

National School 53
navigation 67-68, 70, 99
Nelson, Arthur, the clown 15, 77, 79-84, 86-88, 90-91, 101-102, 104
Norfolk Chronicle 13-14, 20, 31, 34-36, 46, 48, 51, 56, 64, 66, 69, 72, 74-75, 89-90, 116
North-end 52
North Quay 58, 66, 76, 105, 117, 118
Norwich 10, 14, 15, 19-20, 24, 26, 30, 33-34, 36, 45, 47-48, 54, 57, 60, 63, 65-67, 70-71, 74-76, 82, 87, 91, 99-101, 104, 115
Norwich Arms 19, 20

Palmer Geworge Danby 63, 85
Palmer, Samuel Thurell 20, 42-44
Palmer, Charles John 53, 104
Palmer, F. 65
Palmer, John Danby 63

Palmer, W. H. 86
Pier and Haven Commissioners 46
Political Economy 10-11, 56
Preston, John 12, 14, 63, 108, 110, 116
Pullyn, P. (Mayor) 50

Railway bill 67
Reeves, Maud Pember 23
rows 21, 23, 25-26 49, 50, 52
Royal Pavilion, Brighton 78
Runham 18, 58, 61, 65, 105, 112, 117, 119

Scholes, John Joseph 60
Smith, Adam 11, 56
Smith, William 20
Social Toryism 9
South Denes 12
Springfield, T.O. 67
Staff, Julie 6-7, 94, 103
Stephenson, George 45, 62, 65, 69-70
Stephenson, Robert 62
St Nicholas church 28, 52-53

Tay Railway Bridge disaster 93-94
Teasdale 87, 88
The Mill on the Floss 65
Thorndike, J. B. 42
Thurston, William and Elizabeth 27
Tolver, Samuel 44
Tootal, Mr. 68
Town and Haven Commissioners 68
Turnpike 58, 60-61, 63, 105, 111-112
Twist the clown 80-82
Tyndale, Charles 70

Utilitarianism 10, 52, 56, 121
Utting, Louisa 19
Utting, Sarah 42

Vaughan, James Stuart 15
Vauxhall Gardens 16, 102, 117
Vincent, Maria 24

Walker, James 46, 97
Water Frolic 12
Watson, Charlotte 23
Watts, Thomas 33
Wensum, River 66
White, Ian Nimmo 93, 94
Winter, Cornelius Jansen Walter 96

Yare, River 14, 16, 66, 68
Yarmouth and Norwich Railway Company 45, 48, 65-66, 74
Yarmouth Roads 34